# EMERGING
# TECHNOLOGIES

edited by Nora Savage, Michael Gorman, and Anita Street

# EMERGING TECHNOLOGIES

## SOCIO-BEHAVIORAL LIFE CYCLE APPROACHES

PAN STANFORD PUBLISHING

*Published by*

Pan Stanford Publishing Pte. Ltd.
Penthouse Level, Suntec Tower 3
8 Temasek Boulevard
Singapore 038988

Email: editorial@panstanford.com
Web: www.panstanford.com

**British Library Cataloguing-in-Publication Data**
A catalogue record for this book is available from the British Library.

**Emerging Technologies: Socio-Behavioral Life Cycle Approaches**



ISBN 978-981-4411-00-4 (Hardcover)
ISBN 978-981-4411-01-1 (eBook)

Printed in the USA

*The editors would like to dedicate this book to all who work together to develop and manage sustainable socio-technical-natural systems that improve global quality of life.*

# Contents

# Preface

Historically speaking, conventional approaches to life cycle analysis (LCA) methodologies have been ostensibly concerned with public health and environmental impacts from materials, consumer products, processes or activities (of the anthropogenic kind.). The last three decades have brought a litany of continuous developmental improvements within the practice. This is evidenced by a growing interest globally as well as an intensity of awareness that is leading practitioners and researchers to apply LCA, or social life cycle analysis (SLCA), in heretofore very unusual and creative ways. Guinée *et al.* in 2011 in their research paper "Life Cycle Assessment: Past, Present, and Future [*Environ. Sci. Technol.*, 2011, 45(1), 90–96]" observed that on the international front, LCA "is booming" in a multitude of ways that is leading us to the building of a deeper and broader discipline. Some of these more creative uses include studies building codes and materials, military systems, waste incineration and tourism. The creation of international standards and the 2009 work of the European Commission (among many others) that established a life cycle sustainability analysis is a testament and acknowledgement of the legitimacy and credibility the field continues to garner.

As researchers, decision makers, and practitioners, who have each long been involved in LCA on different fronts, we were each very keen to examine the proposition of a more adaptive framework. One that, in part, unapologetically asks the questions of the producers of an emerging technology or material: who will use it; how will it be used; and who will be the primary beneficiaries? The result is a "collaboration of the willing" that includes illuminating submissions from researchers from multiple areas of science and social science including the fields of molecular toxicology, anthropology, behavioral and environmental economics, political economy, ethics, public engagement, technology assessment, governance, etc. The underlying premise of the book is that a more effective approach towards life cycle considerations for emerging materials and technologies should include a wider range of perspectives and disciplines. Ultimately, we hope to inspire new areas of research and deeper dives into existing areas that will shed new light.

Given the broadening application of SLCA, we have included critical analyses on the assessment of emerging technology, responsible innovation and reframing cradle-to-grave approaches for example. In addition, we have included chapters on the effectiveness, as well as the acceptance or adoption of a technology, the dependencies on the prevalent context, materials, supply chain, governance structures, and societal norms or cultural behaviors.

We propose that this framework could be done in three stages with examples from the nanosciences. The book includes critical discussions about trading zones, ethics, behavioral nanotechnology, governance, risk, green design, and urban mining as potential fodder to develop tools for practitioners and decision makers.

As the editors, we have attempted to assemble a volume comprising research and thought pieces of some of today's leading experts on the subject. It is the hope of the editors and contributors that this book will meaningfully contribute to an international dialogue suggesting that a need still exists to further refine and develop holistic approaches that incorporate societal and behavioral dimensions—at the heart of the human-environment/technology interface. The research presented here serves as a proverbial call to arms that will hopefully inspire, provoke and encourage discussion and research. In the first two chapters, we have also created two scenarios to tie together overarching themes that are woven throughout the volume. Each is broken into two parts, the first reflecting the design phase of the lifecycle and the second reflecting what happens when the innovation is introduced into the socio-technical system. These vignettes are placed right after the introduction to stimulate imagination about the social-behavioral-ethical aspects of LCA.

The editors and authors sincerely look forward to more holistic assessments of emerging materials and the resultant improved protection of public health and the environment. The approaches presented in this volume serve merely as examples; further exploration, additional ideas, and adaptions of these ideas are welcome.

As the Noble Prize–winning physicist Werner Heisenberg once said, "The existing scientific concepts cover always only a very limited part of reality, and the other part that has not yet been understood is infinite." These words never rang so true.

**Nora Savage, Michael Gorman, and Anita Street**
Spring 2013

# Introduction

As technology and innovation continue to advance, and as novel materials and devices are developed to improve the quality of life, a better and more thorough assessment of the impacts will be necessary. Life cycle analysis should not only raise key questions about where and when during the product life cycle material and energy impacts could arise but also attempt to ascertain whose quality of life the product might improve or reduce, and why. In the era of increasing populations, decreasing resources, and concentration of wealth in the hands of a decreasing subset of the population, society can no longer afford to ignore or dismiss such questions.

One approach for obtaining answers to these questions and for attaining more accurate assessments of potential impacts is to incorporate behavior, ethics, and philosophy along with other components of social science into life cycle analyses of emerging technologies and products. The figure below, which shows a traditional life cycle analysis, illustrates this point by placing material and energy flows at the top of the triangle. Behavioral and social components and also emerging technological alternatives occupy the base of the triangle. The incorporation of these additional factors cannot be simply off-hand or by-the-way. Careful consideration of the product or technology along the life cycle and the exploration of cultural mores and ethics surrounding various activities are required. Dialogue among the different concerned parties is critical. Communication among them as key parties to the activity or users of the product rather than lectures is important. Phrases such as "If only the public understood the technology they would embrace it," can be appropriately countered with "If only the technologists understood the cultural ramifications of the product, they would think again." Potential misuse of products, particularly by children, should be considered and altered or new uses of products should be assessed especially where targeted marketing and manufacturing practices promote these uses.

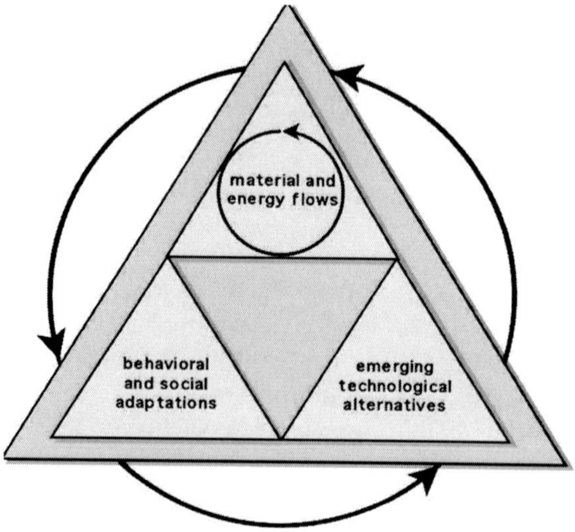


The circles with arrows denote life cycle analyses. The smaller circle around material and energy flows is the usual life cycle analysis done in industrial ecology. The larger circle accompanying the triangle denotes the expanded space explored by this book, a reminder that more traditional forms of life cycle analysis externalize important behavioral and social components, including the kinds of innovations that change the technological options.

The need to incorporate the social sciences along with the physical sciences into any approach toward understanding and assessing potential impacts of emerging materials for life cycle analysis is critical. Social scientists, ethicists and policy-makers raise questions like:

How will users transform the technology?

How will a technology interact with other emerging technologies?

How can governance systems anticipate and manage socio-technical innovations?

Will the new technology result in improvements in the overall global quality of life or merely improvements in the quality of life for a relatively few? Will greater disparities occur between the very rich and the very poor? Increasing this divide will generate social unrest and trigger actions which can adversely impact public health and the environment.

This book offers several options for adopting this more proactive integrated assessment. In addition, two fictional vignettes serve as the backdrop for the importance of these types of approaches.

Chapter 1

# Using Nanotechnology to Filter Water: A Hypothetical Case Study

Michael E. Gorman,[a] Nora F. Savage,[b] and Anita Street[c]

[a] *School of Engineering and Applied Science, University of Virginia, 351 McCormick Road, P.O. Box 400744, Thornton Hall, Charlottesville, VA 22904-4744, USA*

[b] *US EPA, 1200 Pennsylvania Avenue, N.W., Mail Code 8722F, Washington, DC 20460, USA*

[c] *U.S. Department of Energy, Office of Intelligence and Counterintelligence, Science and Technology Division, 1000 Independence Avenue, S.W. - GA-301, Washington, DC 20585, USA*

savage.nora@epa.gov

"Imagine a water filter so fine that no microorganisms or contaminants can get through, and so durable that you can pass the same filter on to your children." Nomian's PowerPoint was pitched to an audience comprised mostly of NGOs and camping equipment firms, but she hoped her slides provided enough details about her fullerene mesh filter to convince even the most skeptical of nanoscientists. Levar Burton had been one of her heroes growing up, and she stole one of his lines from Reading Rainbow: "But you don't have to take my word for it." She pointed to the six prototypes spread across

*Emerging Technologies: Socio-Behavioral Life Cycle Approaches*
Edited by Nora Savage, Michael Gorman, and Anita Street

ISBN 978-981-4411-00-4 (Hardcover), 978-981-4411-01-1 (eBook)
www.panstanford.com

three tables. "Try it yourselves." Next to each filter was water containing fatal doses of *Escherichia coli* and also *Cryptosporidium* and *Giardia*. "I will personally drink any water you pump—just don't pump too much or I will be in the lavatory most of the night!"

Laughter, applause, hands held up for questions but she deliberately stepped off the podium and toward the tables. Her team had tested every prototype but she was a little anxious—any slip-up, and she would feel it first. She touched the antibiotics in her pocket.

People pumped, she drank. "I would bet this filter could turn wine into water," she joked.*

The original plan was to market the filter in the United States for backpackers and other recreational users who wanted the strongest and lightest filter for their use, and invest those revenues in simpler designs for the developing world. Nomian was frustrated with existing backpacking designs because the plastic housing was bulky—she wanted to develop nanocomposites that would reduce its size to about a third—a filter that one could slip in one's pocket with a pen-sized plunger that would move enough water to be useful on a trail.

"The perfect is the enemy of the good," said Feng, one of the angels who financed the work. "But the point of nano is to make breakthroughs. The whole product has to point toward the future."

As Nomian and her team worked on development, nanomaterials and nanotechnology were coming under increasing scrutiny. Nomian was worried that even if they succeeded technologically, they would be blind-sided by changes in regulations.

Meanwhile, people around the world were dying from drinking contaminated water. Nomian needed an alternative to the backpacking design that could be implemented immediately. When she broached this with Feng, her backer and confidante nodded. "Let's focus on those who need this the most."

Feng connected Nomian with an NGO, Potters for Potable Water, that was promoting the use of ceramic filters in the developing world, trying to turn this into a cottage industry. Perhaps the carbon nanotube mesh could be embedded in the ceramic, making a stronger and more effective filter? This product could be developed outside of

---

*Dr Greg Allgood actually drank water he treated at a demo for a State Department Conference, according to Christina Stamper, one of the organizers. (See http://www.csdw.org/csdw/gallery.shtml for a demo of the treatment.)

the United States, avoiding regulatory and testing issues. "I will be the test subject," Nomian told Feng.

Bonding the nanotube mesh to the ceramic was tricky, but Nomian and her team produced several good prototypes. Nomian was the test subject for them in the lab. Because the clay filters relied on gravity rather than a plunger, Nomian added silver nanoparticles to the interior of the filter as an antibacterial agent. If a user became impatient or sloppy and did not properly filter all the contents before use, the silver nanoparticles would reduce the risk.

Nomian took her filters to a village in South Africa where power for all electricity in the village came from a low-capacity diesel generator, and the time the generator could be used depended on the price and availability of fuel. Therefore, the villagers raised and lowered buckets into the well by hand. A test revealed the presence of *E. coli*. The well was open, and the villagers kept cattle. There was no telling what got into the water.

She deliberately took two versions of the filter, one made out of plastic (someday it would be made with nanocomposites) and the other made out of clay by Potters for Potable Water. Nomian stayed in the village and inspected the filters after each use, instructing the villagers on use and maintenance. She relied heavily on Tienz, a civil engineer trained in the United States and who had returned to help improve sanitation and water in his native country. He worked with Nokthula, a villager they paid to help them, to turn lessons about filter use into pictures, and a play that illustrated how to use them properly.

The carbon mesh became partly detached on one of the pottery filters but Nomian saw it right away, made an example of it for the villagers, and removed it. Another pottery filter was dropped and it broke. The villagers preferred the modern pump model anyway.

She and Tienz took a breather with a beer under a tree. "You sure I don't need to put this through the filter?" she laughed.

"This beer is weak because we drink a lot of it—better than the water before you came along. Would prefer it were a cold beer."

Nomian laughed, and for an instant, felt the stress of managing all of this leave her shoulders and back. What a great adventure —

Nokthula was walking toward them, waving her arms. Nomian was delighted; she saw Nokthula as a younger version of herself, and had persuaded her to become a paid member of the project team. Perhaps someday she would become a scientist, or engineer —

"Two of the children are sick. I have been using the filter."

The sun slammed them as they left the shade.

The two children, four-year-old twins, were vomiting and having diarrhea. Tienz took their temperatures—no fever. The mother was sure the filters were the source of the problem. Nomian noted that she was using the pressurized filter, and asked to inspect it. Inside, she saw a white residue, smelled it.

"What's this?"

"It is medicine we got when the clinic came to the village. I thought the filter would keep it clean."

Nomian sighed. "The filters do not work for medicine. You have to refrigerate."

The mother looked down. Tienz shook his head, "She knows. There is not enough power in this village for refrigeration."

"What was the medicine for?"

"Their ears—the traveling doctor said they were infected. This was the last dose."

Tienz took temperatures and looked in the childrens' ears. "Good news—no fever. They need to vomit the rest of the medicine, then we will rinse out their mouths and hydrate them using filtered water. I have got a powder that should help with any further stomach problems."

Nomian was afraid the villagers would reject the pump filter after this, but Tienz and Nokthula created a picture of children getting sick when anything but water was run through the filter. Nomian and Tienz devoted more time to observing the filters in use, but one morning, one of the six filters was missing.

When the filter could not be found on their daily check, Nokthula took them aside.

"The filters will all disappear soon if you do not promise that you will give them to the villagers. I will make sure they are used correctly, and that we collect data."

"But—but these are prototypes!" Nomian said. "I have tested them in the laboratory, but not here, not under these conditions over a long period of time. What will you do if the pump breaks, or the plastic cracks, or the mesh comes off—or someone else from the village uses it in a way we cannot anticipate?"

"You will make us more."

## Acknowledgment

The authors would like to thank Christina Stamper for her assistance

## References

Fauss, E., Gorman, M. E., and Swami, N. (2011). Case-study of an emerging nanotechnology: Identifying environmental risks from silver nanotechnology through an expert elicitation methodology. In S. Ripp & T.B. Henry (Eds.), *Biotechnology and nanotechnology risk assessment: Minding and managing the potential threats around us* (pp. 17–40). Washington, DC: American Chemical Society.,

Chapter 2

# Nanotechnology for Human Health

**Michael E. Gorman,[a] Nora F. Savage,[b] and Anita Street[c]**

[a]*School of Engineering and Applied Science, University of Virginia, 351 McCormick Road, P.O. Box 400744, Thornton Hall, Charlottesville, VA 22904-4744, USA*

[b]*US EPA, 1200 Pennsylvania Avenue, N.W., Mail Code 8722F, Washington, DC 20460, USA*

[c]*U.S. Department of Energy, Office of Intelligence and Counterintelligence, Science and Technology Division, 1000 Independence Avenue, S.W. - GA-301, Washington, DC 20585, USA*

meg3c@virginia.edu

"Today, I am pleased to announce a breakthrough that takes us closer to an end for arteriosclerosis, one of the leading causes of heart disease and stroke—made possible by the support and courage of most of you in this room. "

"Here is the image of a coronary artery in a dog on the verge of a heart attack. And now, right under it—same dog, same artery 24 hours later." There were gasps of amazement from a couple of those in the room that were used to looking at such images.

"We have done this procedure successfully on six dogs."

"Our basic research was supported by the National Institutes of Health (NIH), with additional support from Defense Advanced

*Emerging Technologies: Socio-Behavioral Life Cycle Approaches*
Edited by Nora Savage, Michael Gorman, and Anita Street

ISBN 978-981-4411-00-4 (Hardcover), 978-981-4411-01-1 (eBook)
www.panstanford.com

Research Projects Agency (DARPA)." She made brief eye contact with Amy from NIH and Stu from DARPA. "I would venture to say that this is proof positive that federal agencies *can* work together." (*Laughter*) "Of course, nothing I say here represents the views of either of these agencies. Many of you here have read our initial publications. Since then, as most of you know, we have started a company—Intellimune. Our current goal is to turn the coronary arterial buildup into a target for the body's own immune system, which would create the possibility of a lifelong cleansing mechanism. This study is a first step in that direction. Nanoscale imaging agents are bonded to the fat deposits on the arterial wall, and the white blood cells attack these agents effectively destroying these cells and preventing plaque build-up that can lead to coronary disease."

"The next image is from a camera on a catheter on one section of the dog's artery, showing changes in time-lapse format, at the rate of one image an hour." There were a few appreciative murmurs from those who could see the small changes. "I know these changes may appear to be barely perceptible and hard to see, but look here," she pointed to a part of the screen. "This slightly darker area here is the imaging agent—you can see how the artery widens there as time progresses." More murmurs and a few gasps. "Good—enough of them see it."

"We have developed methods for testing, iterating, and improving these techniques *ex vivo* which means we can limit animal testing to a minimum." Not just a nod toward PETA—Francesca meant it, she hated having to sacrifice animals. Someday, she would develop biological/mechanical hybrid systems specifically for testing.

"This technology is potentially a giant step toward a cure for arteriosclerosis, but it is also provides a promising proof of concept that we can achieve, collectively, something even larger—a technological complement to enhance our existing immune system, under conscious, rational control. The immune system is the result of millions of years of evolution, and has a built-in intelligence. But intelligence itself is also the product of millions of years of evolution. Intelligence can respond more rapidly to new threats to the human body than evolution can. Just think if we could eliminate auto-immune diseases, for example, through conscious modulation and adaptation of the immune system." She was careful to avoid the word control—and also careful not to use the word immortality. Otherwise every religious group in the country would be down her throat. This immune system idea was radical enough.

"Questions for Doctor Steiner?"

"Are you ready to conduct human tests of your nano-enhanced immune system?" Perfect.

"Thank you for asking me that question in front of the Food and Drug Administration (FDA)." She gestured toward Carlos, the FDA representative in the room. (*Laughter*) This was good. "I am pleased to announce we are planning the first human trial." Here she saw Carlos frown and sit up straight. "In the great tradition of Barry Marshall* and other pioneers of medicine, I will be the subject. In the great tradition of Barry Marshall, I might end up with a Nobel, too."

Three months after her press conference, the top YouTube video was the news release from Francesca Steiner's physician at NIH. "Dr. Steiner's immune system reacted to a few of the machines. The surgeon took the necessary steps to flush them out of the system, but the immune response interfered with the controls and a few remained, though they were completely inactive. We are waiting for them to flush out of the system. Here is Dr. Steiner."

The camera moved to Francesca in a hospital bed, propped up and pale. She spoke into a microphone, barely above a whisper. "I am ill not because my nanomachines did not work. The parts of my coronary arteries they worked on are clean. I am ill because my primitive immune system is fighting its ally. This is why we need to couple the natural and technological systems.

"I was only six when an American landed on the moon—and I will never forget it. I want America to aspire to something much greater. I propose a National Initiative on Human Longevity. This initiative would also advance other goals—like space exploration. Imagine not just robots but astronauts with special adaptations to function normally in space. Think of the jobs this initiative will produce. Think of the technological leadership it will give our country. Think of the way we will build a better future for this generation and all that will follow. Think of the huge economic benefits we will reap." She closed her eyes for a moment, rasped: "I ask that every possible lesson be

*Barry Marshall was sure *Helicobacter pylori* caused ulcers, and to test his theory, ingested the organism—which resulted in ulcer symptoms within a few days. He and Robin Warren received the 2005 Nobel Prize in medicine for their work on *H. pylori* and ulcers, which led to successful antibiotic treatments.

extracted from my body if I die. This is not a setback—it is a chance to learn."

-------

Ann could see the image of her own body and the rope, and the beat told her when to skip. She kept missing and a soothing voice said, "Start again." She sighed. Something large and black caught her attention and she shifted her eyes to the street. A van was pulling up in front of the house. She blinked twice to turn off the program. Maybe it was a package!

A man in a uniform got out of the front, went to the back door on her side, opened it and pulled out a ramp. A very big package? Sometimes Grandma sent her presents, like the saddle that turned into a horse when she used her glasses, but her mother always sold them. Mama had gone to the Church to do something. Maybe she could open it and play with it just for a minute—

Out came a chair that stood up as it went through the door and down the ramp. Steel hair, smouldering eyes. Grandma! There were wires sticking out of her arms. Ann was a little frightened, but hoped Grandma was bringing the present instead of sending it.

Grandma rolled up to her so quickly that Ann stepped back.

"Oh, I'm sorry—this thing is so much quicker than walking. I forget. Ann, come take a ride with me, I want to talk to you—and I have something for you at the Institute."

"Mama told me to play here until she got back. She is coming back soon."

"I am your grandmother, child—I will bring you back quickly."

"No, Mama will be very mad at me." Grandma sighed, and for just an instant her steel face looked sad.

"Yes, she will be very mad at me, too—she has been mad at me for years. This body is dying, child, and I wanted to offer you something I can only wish for—the chance to live over a century—perhaps two centuries—perhaps longer at the rate we are making progress. I wanted to show you this future and invite you to come work at the Institute someday, when you are a grown up." The smouldering eyes were on fire now, and Ann felt them bore into her.

"But—but Mama says it isn't natural, and that people died —"

The eyes looked away for an instant, then fixing on her again. "Yes, there were pioneers who volunteered to take risks, like the explorers who tried to start the first Mars colony. They died knowing they were opening up a new world for future generations to experience. So did those who volunteered at the Institute—like me. I never asked anyone to do anything I wouldn't do to myself. And all the volunteers lived longer than they would have—"

Screech of brakes and a car almost bumped the limo. When Ann rode with her friends the car braked automatically, but Mama said they could not afford it.

"Francesca, what are you doing here?" Ann wondered why Mama called Grandma by her first name. Mama's eyes were hotter than Grandma's, and she was moving as fast as the wheelchair.

The man in the gray uniform tried to stand between Mama and Grandma.

"Alaric, this is family. Please wait in the car, thank you." Grandma suddenly looked sad.

"Ann, go into the house, I have to talk to Grandma."

Ann went inside, but she felt sorry for Grandma in the wheelchair with all the wires sticking out of her. She stood in a shadow and looked through the screen door.

"You came knowing I would not be here, didn't you? Didn't you!" Grandma looked down.

"Yes."

"You were going to take her to the Institute, weren't you?"

"Yes, I wanted her to understand—"

"How you ended up in a wheelchair with tubes and wires? You are my mother—I still love you—but is it really you, anymore?"

"I was the guinea pig for so much of this. It will be different for others—it is already different—"

"Did it ever occur to you to ask me if I wanted a mother who was no longer human?"

# Chapter 3

# Assessing Emerging Technology Systems: Why LCA Fails

**Braden R. Allenby***
*Center for Earth Systems Engineering and Management, School of Sustainable Engineering and the Built Environment in Civil Engineering, Arizona State University, Tempe, Arizona, USA*
Braden.Allenby@asu.edu

## Introduction

Life cycle assessment (LCA) methods are an increasingly powerful toolbox used by many practitioners to evaluate the environmental implications of products and materials. The different techniques range from highly quantitative and complicated individualized analyses to online tools of various types generally operating in either top-down or bottom-up mode to semi-qualitative matrix techniques that are useful for streamlined or scoping LCAs (Allenby,

*Braden R. Allenby is the Lincoln Professor of Engineering and Ethics; a President's Professor of Civil, Environmental, and Sustainable Engineering; a Professor of Law; Founding Chair of the Consortium for Emerging Technologies, Military Operations, and National Security; and the Founding Director of the Center for Earth Systems Engineering and Management, at Arizona State University in Tempe, Arizona.

*Emerging Technologies: Socio-Behavioral Life Cycle Approaches*
Edited by Nora Savage, Michael Gorman, and Anita Street

ISBN 978-981-4411-00-4 (Hardcover), 978-981-4411-01-1 (eBook)
www.panstanford.com

2012; Graedel & Allenby 2010). Currently, efforts are underway to expand LCA techniques to include social science dimensions, with the expressed desire to create "sustainability life cycle assessments," or SLCAs (see chapter 1, this volume). While this research will no doubt prove interesting and valuable in extending the current limited remit of LCAs and may be applicable to particular physical artifacts or uses of materials, it is highly unlikely that such approaches will be useful in evaluating the implications of emerging technology systems for a number of fundamental reasons. First, powerful emerging technology systems are inherently unpredictable and cause profound and unforeseeable changes across economic, social, institutional, and cultural systems (Freeman & Louca, 2001; Rosenberg and Birdzell, 1986); they are thus not well characterized by techniques that rely on historical data. Second, technology systems, as opposed to mere artifacts or particular uses of given materials, are quintessential complex adaptive systems. This means that any coherent modeling effort, such as an LCA, is necessarily at best only partial (Allenby & Sarewitz, 2011). Finally, and perhaps most fundamentally, technology systems do not have life cycles: what is the life cycle of railroad technology (as opposed to, say, the life cycle of a particular locomotive, which can be defined and measured)? The life cycle of the Internet? Of social networking? Of pharmacological cognitive enhancement technology? This does not mean that analytical frameworks cannot be deployed; indeed, one possibility is presented in this chapter. But it does mean that LCA, in any recognizable configuration, is not an appropriate method for understanding emerging technology systems.

Before engaging in a more detailed discussion, it is worth emphasizing the difference between technology as artifact and emerging technology systems. "Technology" for many people is a physical object—a car, a refrigerator, a smart phone. And if one is interested in the immediate environmental impacts over the life cycle of a particular artifact, LCA is a useful way to structure a scenario to enhance analysis. Indeed, techniques such as Carnegie Mellon University's EIO-LCA model (which can be found at www.eiolca.net) enable one to model environmental impacts of particular families of artifacts across the economy as a whole. Similarly, one can use LCA techniques to explore the choice of solvent for a particular application, or the environmental impacts of using a particular metal in an automobile design. Streamlined LCA techniques can enable one to view and contrast sets of technologies; indeed, AT&T used such a matrix LCA tool to evaluate environmental considerations across a number of its products in the early 1990s (Graedel & Allenby, 1995).

But in the case of emerging technologies, one often does not have much indication of where technological evolution will lead, much less a stabilized design. Moreover, these technology systems are very potent, and their social, cultural, and institutional implications—not to mention their environmental implications—are often far more important and meaningful than any environmental considerations identified by an LCA approach (Allenby & Sarewitz, 2011; Bijker *et al.*, 1997; Grubler, 1998). Thus, for example, one might be able to do (a modified) LCA on a vaccine, although the human dimensions of such a technology tend to outweigh the environmental considerations involved (if a vaccine would save your child, are you going to refuse it because it contributes to a small degree to global climate change?). But an LCA of "vaccine technologies" is simply irrelevant: it is to apply a tool designed for an apple to a fundamentally different orange.

## Background: LCA Methods and Complex Systems

The other chapters in this volume provide all the relevant detail regarding LCA, and integration of social science into LCA methodologies, that the average reader should require. Against this background, it is useful to the discussion in this chapter to identify the three categories of important limitations regarding LCA methodologies and how they impact application of any such analytical tools to complex technology systems, especially those that are just beginning to emerge. The second section will focus on explicating the characteristics of such systems, using railroad technology as a case study. Finally a framework for evaluating technologies based on work I have performed with D. Sarewitz that is not based on classic life cycle analysis, but rather on the relationship of the technology to society and culture, will be suggested (Allenby & Sarewitz, 2011).

The important LCA categories are implicit and explicit boundaries, critical assumptions, and data issues. These categories, although not always identified as such, are familiar to most practitioners, and need not be explored in detail here (Allenby, 2012; Graedel & Allenby, 2010; Hendrickson *et al.*).

The most important boundary condition from the viewpoint of complex adaptive technology systems, including emergent ones, is the one upon which this volume focuses, the environmental bias of all LCA methodologies. Indeed, this is true of the broader field of industrial ecology, within which LCA is a particular set of methods. Although there are continuing efforts to broaden the scope of industrial ecology and its various toolboxes, in the continuing emphasis on environmental considerations one can detect the social and policy origins of much of industrial ecology and its practices. This is not only because many of the people who originally worked on industrial ecology, and on LCA, came from environmental backgrounds; it is also because much of the original problem definition, and the focus of activism and regulation, was centered on such issues (Allenby, 2012).

A second, more subtle bias in industrial ecology and LCA methods is the concentration on manufacturing sectors and material and energy flows. This also reflects the evolutionary history of these fields, in that manufacturing was the obvious locus of environmental activism during the period when these areas of study developed, and manufacturing firms were the ones most affected by environmental regulations and politics at that point. This history helps explain two other implicit boundaries to industrial ecology and LCA: the emphasis on material and energy flows rather than information systems and, more broadly, services; and a general tendency to prioritize environmental issues important to advanced economies rather than those important to developing economies (a boundary that is being renegotiated as newly industrializing countries such as the BRICs (Brazil, Russia, India, and China) become manufacturing and economic powers).

Not surprisingly, critical assumptions underlying both industrial ecology and LCA methodologies of various kinds tend to reflect the same biases. Thus, for example, both industrial ecology generally and LCA in particular prioritize environmental values as above others, exemplifying Aldo Leopold's famous land ethic that a thing is right when it supports ecosystems, and wrong otherwise (Leopold, 1949). This assumption lies at the heart of both the field of industrial ecology and LCA methods, and it forms a powerful, if little appreciated, barrier to expanding LCA to include other values, unless those values are subordinated to environmental ones. It is reflected, for example, in the strong tendency to define "sustainable engineering" in terms of "green engineering," which despite best efforts of practitioners conflates sustainability with environmental values (see, e.g., Allenby, 2012; Graedel & Allenby, 2010). More fundamentally, it raises an ontological issue regarding LCA: is any LCA methodology which does not elevate environmental values above all others still an LCA methodology? And if it does not, isn't any "sustainable LCA" simply a restatement of normative environmental activism? Such questions are difficult to answer a priori, but they raise deep and underappreciated challenges to the industrial ecology and LCA communities; moreover, they also raise serious questions about extending LCA to technology systems, because such systems inevitably raise serious questions across all domains, not just the environmental.

More subtle but equally important is the assumption that any internally coherent model or method, be it LCA or any other, is adequate to describe a complex adaptive system such as an emerging technology system. This assumption is, unfortunately, wrong. A model, after all, is a mechanism that simplifies reality by using a coherent set of assumptions and rules to determine what should be included in the model, and what can be excluded. But in doing so, it necessarily translates a complex adaptive system into a simpler structure—which is fine so long as everyone understands what is going on, and uses the results of the model accordingly. But it fails when—as with emerging technology systems—it is both unclear a priori what "matters" for purposes of understanding the system, and the act of simplification necessarily strips out information that is at least potentially meaningful. Sometimes complex systems simply can't be boiled down into bumper sticker language, because the meaning lies in the complexity itself (Allenby, 2012; Allenby & Sarewitz, 2011). It should not be thought that this represents only an academic caveat, either: a major reason for the climate change policy train wreck is precisely the unfortunate tendency of activists and climate change scientists to assume that an environmental perspective should dominate all other elements of the very complex adaptive system that climate change engages. Indeed, this is a major problem with geoengineering proposals as well: although they come from a different community than the one that generally engages with climate change activism, they reflect the same simple system perspective. To deploy a technology system that by definition is potent enough to dramatically affect climate physics and chemistry for solely environmental reasons (i.e., to reduce climate change forcing) shows a dangerous misunderstanding of both complexity and technology systems in general (Allenby, 2011).

The data issues that bedevil LCA—incomplete and outdated data, difficulty in ensuring comparability across data sources, local versus averaged data, and so forth—are well known to practitioners, who generally use the results of their studies with appropriate caution. But emerging technology systems raise an entirely new set of data issues: it is not just that data are difficult to get, but, worse, data are even in theory unobtainable. The major factor here is simply one of predictability: LCA techniques implicitly assume that historical data will remain stable enough over the period under analysis so that they can provide guidance to future behavior: if a region gets most of its power from coal-fired facilities, it is reasonable to assume that production of a widget in the immediate future will necessarily involve similar energy consumption patterns. But with an emerging technology of more than trivial import, it is precisely the unpredictability of future evolutionary paths that is of most interest: historical data by definition are inadequate. For example, if one lived in the early 1800s and were to evaluate the environmental impacts of a locomotive sitting on its rails, the obvious points would have been, for instance, fuel and water consumption and local emissions and noise impacts. One would not have immediately understood that railroad technology would result in most of the American Midwest being converted from wetlands and flood plain ecosystem to massive industrial agriculture landscapes, with concomitant biodiversity and resource impacts. Nor would one have understood that enabling food production at industrial scale would in turn enable population growth around the world, which in turn would have incalculable effects on environments at all scales. It is only in retrospect, after such technologies have been deployed, that one can see how all the interacting factors—institutional, political, economic, technological, social, institutional, and cultural—played out.

This point deserves emphasis: it is not that we don't have the data to do LCAs, or any other methodologies. It is that we *can't* have the data until it is revealed in real time—and often, because of the complexity of these systems and various psychological and cultural barriers to perceiving change as it is actually occurring, we can't have the data until a significant period of time has passed. Put another way, any method that assumes validity of historical data is invalid when applied to evolutionary processes involving complex adaptive systems. It does not mean we can't know or develop rational policies, but it does mean that LCAs, or any similar methodologies, fail in the case of emerging technology systems. Before presenting one framework that brings some light to this admittedly difficult dilemma, however, it is worth exploring technology systems in more depth so that their inherent unpredictability and uncertainty can be understood.

## Technology Systems: The Railroad Example

Many engineers are familiar with complicated systems, with lots of parts and interconnections; they also study dynamic complexity, which is the complex behavior that arises as systems, complicated or in some cases apparently simple, move through time. But emerging technologies engage two additional levels of complexity: "wicked" complexity and earth systems complexity (Allenby & Sarewitz, 2011). Emerging technology systems are not just a collection of artifacts but integral parts of the human/natural/built systems that characterize the Anthropocene, the present "Age of Humans" (*Nature*, 2003). Technologies at this scale not only physically construct a human Earth, but they do so by coupling natural systems to human and social systems, which have a far different, and higher, degree of complexity—"wicked" complexity (Bijker *et al.*, 1997; Heidegger, 1997; Rittel & Webber, 1973).

Understanding technology systems in this way aligns with the concept of what economic historians call "long waves" of innovation, or "Kondratiev waves," after the Russian economist Nikolai Kondratiev. Kondratiev waves are long-term patterns of innovation that develop around core fundamental technologies, with each wave accompanied by additional technological, economic, political, cultural, social, and institutional changes (Freeman & Louca, 2001). Thus, although the dates and choices of core technology are not etched in stone, one can identify "technology clusters" that characterize different periods of institutional and technological innovation. Railroads and steam technology powered a wave from about 1840 to 1890, with steel, heavy engineering, and electricity characterizing a wave from about 1890 to 1930. Subsequently, automobile, petroleum, and aircraft technologies created a mass consumption wave from about

1930 to 1990; the current wave is sometimes characterized as that of the "Five Horsemen"—nanotechnology, biotechnology, information and communication technology (ICT), robotics, and applied cognitive science (Allenby, 2012). Each of these technology clusters coevolved with unpredictable and profound institutional, organizational, economic, cultural, and political changes (Freeman & Louca, 2001 Rosenberg & Birdzell, 1986). So, for example, the specialization and professionalization of managerial systems began during the railroad wave and spread throughout the economy during the heavy industry cluster—which, of course, could not have evolved until a technology like the railroad created an infrastructure that enabled economies of scale across national regions (including previously less accessible internal regions such as the Canadian and American prairies). Mass market consumerism was not conceived of until mass production, especially of automobiles and large consumer goods such as washing machines and refrigerators, developed—and the financial innovation of consumer credit made those products accessible to the masses. It wasn't until the development of modern ICT that the far more networked, flexible structure characterizing current production systems—and the financial infrastructure that supports them—began to evolve (Castells, 2000). These technology clusters are not just artifacts and a little social change: they represent fundamental shifts to new metastable states across all Earth systems: human, built, and natural.

A brief discussion of a familiar technology, railroads, may provide deeper insight. This technology system is quite familiar today, but as it first began its rapid expansion in the early to middle 1800s, it was both new and impressive statement of human potential (for good or evil, depending on the commenter). Although not appreciated at the time, it was also the end of an era: the world after the railroad was profoundly different than the one existing before the railroad. The differences were fundamental, linked to each other, unpredictable, and extended across social, cultural, natural, economic, institutional, and other systems. It should be understood that the railroad did not "cause" these perturbations, but, rather, it *coevolved* with them; causality is a property of simple systems, and extending it to complex systems often becomes a form of determinism (i.e., "technological determinism" would be the argument that the technology "caused" the accompanying changes, rather than the understanding that the constellation of changes accompanying a new Kondratiev wave coevolved together).

It is difficult to overstate the environmental implications of railroad technology. Certainly the material and energy consumption of the new technology, and related emissions, which would have been the focus of a contemporary LCA, existed. But in terms of the technology system, these effects were vanishingly small. Rather, it is the systemic impacts which dominate: railroads created a robust transportation infrastructure, cut time in shipment dramatically (thus enabling global distribution of grain, for example), and enabled economic access to continental interiors. This in turn enabled the growth of industrial agriculture, with the end result of completely changing regional ecologies in many areas. In the American Midwest, for example, railroads transformed the landscape dramatically because it meant grain could be grown anywhere a railroad feeder line existed, fed into Chicago, consolidated in Chicago and sent to the port of New York City, from where steam ships could carry it around the world. Previous technology could not economically transport a perishable bulk cargo such as grain more than a few miles overland; now, farmers throughout the middle of North America fed a global market (Cronon, 1991). Note also that to the extent railroads enabled urban centers such as Chicago to develop, they share responsibility for the concomitant environmental changes to which such cities contributed.

But in many senses the environmental impacts of railroads, consequent as they may have been, were epiphenomenon piggybacking on more fundamental social, cultural, and institutional change. For example, core technologies in a Kondratiev wave change other technology systems. In the case of railroads, one is constructing an integrated network that requires coextensive management structures (not unlike, e.g., the management networks built into computer chips). In particular, two new technology networks must coevolve: a signaling network, so that information and material movements throughout the network may be coordinated, and a timing function, so that movements across the network may be located in time and space in relation to each other. In the case of the railroad, telegraph technology (often laid along the same lines) coevolved to perform the signaling function, assuming the role of necessary coordination mechanism for regional integrated rail systems (Grubler, 1998).

The network time function is a more complicated story, in part because time is more deeply embedded in cultural and institutional behavior than telegraph technology. It was clear that railroads required a uniform, precise system of time that reflected the speed of the new technology compared to earlier transportation forms—canal boats, sailing ships, and horse and cart routes. Before railroads, local times could be, and were, isolated and idiosyncratic. Thus, Schivelbusch (1977) notes that in the United Kingdom prior to the railroad London time was four minutes ahead of Reading, over seven minutes ahead of Cirencester, and fourteen minutes ahead of Bridgewater. Similar patterns held in the United States, where even in the 1850s there were more than 200 different local times (Beattie, 2009). Moreover, the adaptation to uniform systems of time was not smooth; for a considerable time in the United States, each train company had its own time, so that stations serving several train companies had different clocks [Buffalo had three different clocks at one point, Pittsburgh six, and especially when trains were still relatively slow, the railroad companies worked with multiple internal times (Beattie, 2009)]. By 1883, however, railroad firms in the United States had established the four time zones used today (Eastern, Central, Mountain, and Pacific), although regional standard time did not gain legal recognition in the United States until 1918 (Beattie, 2009; Schivelbusch, 1997). In short, the "industrial time" system that to most moderns is intuitive and virtually invisible was a coevolutionary product of the railroad technology system (Rosenberg and Birdzell, 1986).

There were also more immediate psychological effects of railroad technology arising from the way that "natural" rhythms of transport—those familiar rhythms associated with wind-driven ships, horses, or simply walking, for example—were displaced by a purely technological system that was much faster, stronger, and impervious to the whims of natural systems: muddy roads or lack of wind no longer stopped scheduled transportation. Moreover, such transport meant that the familiar temporal and spatial distance between points was slashed in unfamiliar, and somewhat disturbing, ways, as the words of Heinrich Heine, written in 1843 with the opening of new rail lines across France, illustrate (quoted in Schivelbusch, 1977, at 37):

> What changes must now occur, in our way of looking at things, in our notions! Even the elementary concepts of time and space have

> begun to vacillate. Space is killed by the railways, and we are left with time alone. . . . Now you can travel to Orleans in four and a half hours, and it takes no longer to get to Rouen. Just imagine what will happen when the lines to Belgium and Germany are completed and connected up with their railways! I feel as if the mountains and forests of all countries were advancing on Paris. Even now, I can smell the German linden trees; the North Sea's breakers are rolling against my door.

Passengers were told that traveling at the fantastic speed of 25 miles per hour would kill them, and that at the least they were acting against the obvious will of God: "If God had designed that His intelligent creatures should travel at the frightful speed of 15 miles an hour by steam, He would have foretold it through His holy prophets. It is a device of Satan to lead immortal souls down to Hell." (Ohio School Board, 1828, quoted in Nye, 1994, at 57). More prosaically, passengers complained of being treated like baggage, impersonal packages to be delivered rather than as individuals (Schivelbusch, 1977). The postmodern concerns about fragmentation of time and space, and complaints about impersonal security measures at airports, are not as new as one might think and have roots that can be traced back to the railroads—which is not surprising in retrospect, since this was the first technology system to break people free of "natural" transport modes. More fundamentally, the impact of major technology systems on human perception, concepts of time and space, and what the human is with respect to different environments again cautions against oversimplistic applications of quantitative methodologies that purport to capture the complexities of such systems. An LCA would be hard put to identify and quantify these conditions, much less balance them objectively against environmental considerations.

Railroad technology also had profound economic effects. This is partially because their demand for capital was insatiable. Whereas the early factory system was financially supported by aristocrats, landowners, and factory owners using their own capital (at least in the United Kingdom), such an informal and individualistic financial structure was nowhere near adequate to support the huge capital requirements of railroad firms [railroad construction was the single most important stimulus to industrial growth in Western Europe by the 1840s (Freeman & Louca, 2001)]. Institutionally, these demands led to a more sophisticated financial infrastructure at the level of the economy as a whole. Additionally, the complexity of the institutions that arose to build and manage these systems required new sophistication in management techniques. Early factory capitalism had, per Adam Smith's famous pin factory example, required a division of labor among factory workers, but the factory owner usually did the managing for the factory himself. The scale of railroad enterprises, on the other hand, required a division of white collar labor, leading to differentiated skill sets such as accounting, planning, human resources, and administrative systems, with specialized professionals (Freeman & Louca, 2001). Railroads coevolved with a new, more complex, and more powerful model of industrial capitalism (developments which were paralleled by increasing industrialization in the agricultural sector as well, also coevolving with rail technology, which enabled the necessary scale of transport to support scale economies in agriculture).

At the level of the nation, railroads made possible the scale economies that led to a dramatic restructuring of economic activity. The American economy, for example, was growing during the late 1700s and early 1800s, but was still primarily characterized by rural villages and local economies, in large part because transport between regions, except over waterways, was difficult, expensive, and time-consuming (Beattie, 2009). This economy was completely swept away in the 1800s as the trusts and monopolies made possible by railroad transportation—Big Sugar, Big Tobacco, Big Oil, Big Steel—built national production and distribution systems and operated with prices and political clout that local operators simply could not compete with (Bruchey, 1980). In enabling these national markets and national institutions, railroads fundamentally changed economic and power structures, not just directly validating and solidifying the continental scale of the American state but also indirectly supporting the underlying cultural frameworks behind the State, such as the doctrines of Manifest Destiny and American exceptionalism (the view that America is unique among nations and serves as a guide and beacon to others—a "shining City on the Hill").

A related dimension of this constellation of institutional change that coevolved with the railroad was a major shift of economic power from agriculture to industrial firms accompanied by a more subtle shift in cultural authority (Marx, 1964; Nye, 1994). This latter effect is often not sufficiently appreciated, but the degree to which railroads contributed to a fundamental and radical shift of teleological focus in American culture from Jeffersonian agrarianism, an Edenic teleology, to a technology-driven New Jerusalem, is remarkable (Marx, 1964; Nye, 1994). What is more, this cultural schism continues to the present day, with the sustainability and environmental discourses leaning toward an Edenic teleology, while the industrial, commercial, and science and technology communities tend toward the New Jerusalem teleology (Allenby, 2009). This shift in worldview, where technology is perceived not as a challenge to agrarian Eden but as a means to achieve a high technology New Jerusalem does much to explain the strong embrace of technology in the New World as opposed to, for example, the environmentalist ideology that is so powerful in Europe (Nye, 2003). It thus remains a politically and culturally potent archetype critical to global cultural and power patterns [and not limited to the United States at this point; China, with a leadership dominated by engineers, may currently exemplify the New Jerusalem teleology perhaps better than the United States—though they would not of course call it that (Elvin, 2004)].

The railroad example clarifies several relevant characteristics of the evolution of technology systems. Most important, perhaps, is that the evolutionary paths of such systems are unpredictable, and, equally as important, render contingent many assumptions that are routinely treated as fixed. Thus, for example, the global, and virtually universal, modern time structure that is so familiar that it is essentially never even questioned is neither inherent in the "nature" of things nor historically ancient. It was not the way pre-railroad American agrarian society or European cultures perceived time; it is a product of the demands of a certain kind of technology system. Moreover, it would not have been foreseeable by pre-railroad societies: globally ordered time frames were not just unpredictable a priori, they were difficult to even conceive in social and cultural systems that had neither any need, nor any concept, of unified and ordered temporal frameworks at a planetary scale. Similarly, there were tens of thousands of owners and employees of small businesses and local economic activities who watched as the first train ran through their village or town without realizing that their economic death rode on those rails. They certainly had every incentive to see it coming, but they did not—perceiving the implications of such technology systems founders on wicked complexity. And it was not just that they had not collected the relevant data; it is that the relevant data did not exist—not until the technology began interacting with other old and new technologies, with social and institutional factors, with culture. Predicting the future by looking at data from the past would have been not futile, but seriously misleading. In short, projecting the effects of technology systems before they are actually adopted is not just hard but, given the complexity (especially reflexivity) of the systems, probably impossible.

This is not just a data problem, of course. The Schumpeterian "gale of creative destruction" generated by capitalistic innovation is unpredictable in part because any significant technology destabilizes existing institutions and power relationships and thus, to some degree, cultural assumptions. Accordingly, it is usually opposed by many, in a contest whose outcome cannot be predicted. It is historically true that, even if such opposition is successful, it will probably not halt the evolution of technology—for example, the European Union has been unable to halt the development of genetically modified crops, and the Bush Administration failed to halt stem cell research. Nonetheless, strong opposition to a technology system in a globalized economy where technology and economic power are significant components of cultural power can pass dominance over time to other cultures where opposition is less effective (Kennedy, 1989).

Another important thing to remember about the railroad example is that it probably understates the degree of change that is likely given current conditions and the wave of technological change bearing down on global society for three reasons. First, there are not just one or two technology systems undergoing rapid evolution today, but the Five Horsemen: nanotechnology, biotechnology, robotics, ICT, and applied cognitive science (Allenby, 2012; Garreau, 2004). Moreover, all five of these technology domains are foundational, in the sense that secondary technology systems are built on them. The result is unpredictable, fundamental, and accelerating technological change across the entire technological frontier. Second, the rate of change in technology systems and, accordingly, in coupled social, institutional, cultural, and other systems is already outrunning social and institutional controls, and it continues to accelerate. Under such circumstances, the illusion of control which is helpful for psychological and social stability is increasingly undermined, with indeterminate but potentially significant implications (such as increasing global fundamentalism as individuals unable to keep up with technological change retreat into the psychological security of fundamentalist belief systems). Third, railroads and similar technologies, despite their obvious psychological impacts, were generally external to the human; put another way, the stability of the basic physical and cognitive human structure could be assumed. Today, however, the human itself is becoming a design space, creating an ambiguity between the designed and the designer and a reflexivity between technology and its creator, that is entirely unprecedented (Allenby & Sarewitz, 2011; Stock, 2003).

This does not, however, mean that one cannot develop methods that at least provide some perspective on emerging technologies. Indeed, one schematic for thinking about these systems is discussed in the next section. But, given the uncertainty and unpredictability of the future, it does mean that any methodology based on historical data, or purporting to provide predictive power as opposed to admittedly unrealistic scenarios, is suspect.

## Levels of Technology Systems

The exploration of technology systems* begins with the obvious point that technology is both an artifact and something larger that cannot be separated from the human, but that is in fact integral to being human (Clark, 2003; Heidegger, 1977). A Cro-Magnon unconsciously holding a rock is a very different creature than a Cro-Magnon holding the same rock but recognizing it as a powerful weapon to be used against threats or to obtain food (a point made nicely in the beginning sequence of the film 2001). In both cases the artifact—the rock—is the same, but the shift in context makes a huge difference in implication. The railroad case study makes the point that powerful technology systems cannot be isolated from their broader context: the social, cultural, and institutional framework with which they coevolve. This swirling coevolution of the human, society, and artifact cannot be evaded, and it greatly complicates any sort of simple methodological approach. Taken as a whole, then, such examples suggest that a framework for analysis of technology systems would be well advised to begin by considering not life cycles,

---

*Much of the discussion in this section is drawn from Allenby and Sarewitz, *The Techno-Human Condition*, published in 2012 by the MIT press, where the interested reader can find a far more detailed explanation of both technology and technological systems and the implications of having to consider the human as a design space.

which as the railroad example illustrates is not a category that fits such systems well (if at all), but by considering the complexity of context within which the artifact is embedded. Same rock, different context, creating different analytical results.

It is useful in understanding technology to begin from the simplest point: technologies are, above all, created to accomplish particular tasks or to meet certain goals. They are means by which human intent can be realized via the technology with a very high probability. A watch represents industrial time, social control of individual preferences, and an entire litany of meaning—but it is also a way by which an individual can, with high probability, know what time it is. This simple fact is often forgotten in the swirl of pretentious postmodernist critique which occasionally surrounds the concept of technology. We choose technologies because they help us accomplish what we want to accomplish. A stone in the Cro-Magnon's fist allows him to kill his dinner; a watch on my wrist allows me to tell time; an airplane flies me from one point to another with staggering safety and efficiency. At this stage—call it Level I—a technology becomes the simple means by which an individual interacts with his or her environment in predictable and useful ways.

But of course that is not the only stage of technology. The airplane as artifact may be quite useful and safe, but it is embedded in a social and institutional framework that can be quite frustrating and make the journey from point A to point B less than predictable. One might be pulled aside by security personnel based on secret lists of names; one might not make a connection because takeoff is delayed due to airspace congestion; one might be delayed because of dust storms (at least in Phoenix, where I live). One might live in a small city that no longer generates enough traffic to warrant scheduled service, so that no carrier will fly the reliable, dependable airplane into the airport in the first place. Thus, technologies are both reliable cause–effect machines (Level I), but they are simultaneously components of complex systems within which the functionality of the individual technology may be heavily modified.

This is the Level II category, and it is a familiar one. A car, for example, works perfectly well as a means to get to work (Level I)—but it may end up in a traffic jam, or the relevant road may be closed for construction, or any number of other systemic conditions may impede its easy use (Level II). At Level II, then, the goal is not coextensive with the technology, and the complex system within which the technology is embedded results in emergent behaviors that cannot be predicted from the behavior of individual units at Level I.

But just considering these two levels is inadequate in light of the railroad example. An airplane, for example, is not just part of a complicated transportation infrastructure with its familiar economic, technical, and institutional dimensions (Level II) but also has broader effects as a global transportation network: jet technologies have changed the way epidemics flow around the world; have resulted in significant psychological shifts in the way people view their world (a continuation of the trends toward time and space compression noted with railroads); have devastated ecosystems not with their emissions but by enabling a newly wealthy older cadre of ecotourists who are able to visit, and overwhelm, previously unreachable ecosystems; have generated an area of engineering practice (aerospace engineering) and high-technology manufacturing that in turn generates new technologies with their own, unpredictable implications; have been repurposed by nongovernmental organizations as useful weapons for terrorist attacks, thereby engendering social and cultural responses with significant implications for personal liberty.

These emergent behaviors at regional and global scale, integrating across many different domains and networks, constitute Level III complexity. Or again, the automobile get people from one place to another quite effectively (Level I), but also functions in a much more complicated network: malls, suburbs, highway systems, petroleum delivery infrastructures, road congestion and construction, and so forth. Groups of cars also create emergent behaviors that can at least temporarily subvert the Level I usefulness to the individual of the car as artifact—traffic jams, for example, or demand for gasoline that, together with emergent scarcity in petroleum networks as a result of policy or exigency, leads to higher gas prices. This is classic Level II behavior. But the automobile as the basis of a Kondratiev wave did far more: it coevolved with significant changes in environmental and resource systems; it coevolved with mass-market consumer capitalism and constituent subsystems such as individual credit;

it created entirely new behavioral and aesthetic subcultures and stereotypes (muscle car and drag race cultures and heavy influences in popular music, for example); it created opportunities for, and a sense of, extraordinary human freedom, especially for women who were otherwise trapped in their houses by social and economic patterns. The automobile as an important component of an earth system structure that would not otherwise exist is a Level III system.

Level I tends to be fairly self-evident, but the divisions between Level II and Level III are not necessarily clear. Part of the reason is that any schema is necessarily somewhat arbitrary. But it is also difficult because the same artifact is expressing different behaviors and implications at different levels of complexity simultaneously, and the properties that emerge at Level I, for example, may be totally different, and even mutually exclusive, with those that emerge at Level III. Thus, for example, a series of very expensive new weapons may provide a nation with significant battlefield advantage at Level I, but because they skew investment in the economy toward defense and away from commercial and industrial products, and encourage what Paul Kennedy calls "imperial overstretch," they undermine the long-term stability and security of that nation, leading to its collapse (Kennedy, 1989). Thus, a technology that at Level I supports national security may undermine it at Level III. Lest this be thought a purely academic observation, it is worth noting that the Bush invasion of Iraq may be just such an instance: originally justified in terms of national security in simple Level I terms [prevent use of WMD weapons of mass destruction (WMD) against Israel or the United States], the Level III repercussions may include a permanent loss of American power and status, especially given the economic weakness that ensued.

Nonetheless, there are in practice major perceptual differences between Levels I and II, and Level III. Consider the familiar case of automobiles: a driver knows her car and how it behaves when she drives it (Level I), even while she complains about traffic congestion and the cost of fuel, all the while expecting both infrastructures to be widely and safely available (Level II). She thinks nothing of dropping her child at a day care miles from her house, stopping at a drive-through for coffee and doughnuts on her way to work, and running a series of errands at various shops around town after she gets out of work (also Level II). But few drivers connect their personal vehicle to the wonder of personal credit, the convenient availability of big box retailers selling vast amounts of consumer goods that are sourced and manufactured through global supply networks, and continents conveniently covered with road and highway systems, nor are they aware of the extraordinary degree to which they psychologically construct their automobile to be emblematic of personal freedom (digital natives may similarly regard their social networking technology as a "technology of freedom," even while their less ICT adept elders view it as an increasingly powerful constraint on freedom—the psychological dimension of technology as freedom differs strongly between subcultures). Some may be vaguely aware of connections between fossil fuels and climate change, but they will have little appreciation for the systemic impacts on natural cycles and material flows that are directly and indirectly a result of automotive technology. They cannot be blamed for this, for most scientists and analysts are equally blind to such Level III emergent behaviors.

Note the difficulty this poses for ethical and rational management of emerging technologies. Almost all technologies that are actually deployed will have positive Level I effects, at least to some degree and for some elements of the population—if it were otherwise, they would not be deployed. The conflicts come in because Level II and Level III emergent behaviors do not necessarily track the Level I benefits. It is quite difficult to compare costs and benefits across levels, and doing so without explicitly recognizing that one is comparing different categories can often lead to deep confusion. Moreover, as discussed above, Level II and Level III behaviors are often difficult or impossible to predict, which means that different interests can generate normative hypothetical scenarios which appear to be realistic but in fact may have virtually no predictive power.

One can create a matrix that helps order these effects. Consider the example of a vaccine. If the goal is to reduce the chances that the individual exposed to the technology will get sick from the disease vector for which the vaccine is designed, the goal of the technology and the technology itself are coextensive and, accordingly, the technology is a good one and should be adopted. This is Level I.

Technology Level Matrix: Vaccine Example

| Policy response / Technology level | Goals | Policy response |
|---|---|---|
| Level I | Reduce Individual's risk of specific disease by use of vaccine | Goal and technology coextensive with high probability of success: implement technology |
| Level II | E.g., increase economic growth in developing countries by reducing costs of disease | Implementation of technology may not help meet goal and might even impede progress |
| Level III | Improve human well-being through vaccine technology | What other systems are affected by vaccination programs; how do they respond? |

Let's say, however, that one wants to implement a vaccination program to improve economic performance in a particular nation by creating a healthier work force. Here, most people intuitively sense a connection between the technology and the goal, although it is not straightforward. For example, it is entirely possible that establishing better water and sewage infrastructure could achieve better health for the working population with less expense, or that too much effort expended on a vaccination program, and not enough on supporting a national public health infrastructure, would waste valuable human and financial resources. Even though this is only one goal of many that might lie behind a vaccination program, this engages a set of Level II issues and behaviors, many of which are coupled to each other, resulting in a fairly complex system that would require close attention if it were to have a chance of succeeding. But if the goal is to, for example, improve the quality of life for a nation's citizens through a vaccination program, one begins to detect Level III issues and questions emerging. For example, might a vaccination program contribute to a youth bulge in demographics which, if not educated and employed, creates a pool of disaffected teens which terrorist organizations can radicalize? Does the vaccine control a disease that previously in that society inhibited behaviors viewed as undesirable, thus contributing to a shift in behaviors that the society finds unacceptable and perhaps undercutting social acceptance of vaccine technology in highly desirable cases? These are clearly hypotheticals, which is precisely the point—technology systems will generate Level III effects, but what they may be, even in broad scope, cannot be known until the behavior actually emerges (which, to make things worse, may involve significant lag times in complex adaptive systems, so one cannot perceive and react until substantial damage is already done). The same technology is involved in all three levels, but the analysis and policy responses are quite different: if one considers a vaccine to be a means of reducing disease, it looks like a Level I technology; if one considers a vaccine to be a means of improving economic growth, it looks like a Level II technology; if one considers a vaccine to be a means of raising the quality of life of a society, it becomes part of long-term demographic trends and subsequent political and social evolution and looks like a Level III technology.

## Conclusion

The unpredictability of emerging technology systems, and their coevolutionary relationship to psychological, economic, social, institutional, political, and cultural systems, means that traditional modeling methods of any type are of limited use. This is particularly true where the methods or tools depend on a life cycle structure, because emerging technology systems do not have life cycles in the traditional sense. It is also true where a set of tools or methods reflects the worldviews and values of a specific discourse, such as environmentalism or sustainability, which at best can provide only a partial perspective of a complex adaptive system such as an emerging technology. This does not, however, mean that methods for evaluating such technologies cannot be developed; it does suggest that such methods should be keyed to degrees of complexity of the system.

### References

Allenby, B. R. (2009). The industrial ecology of emerging technologies: Complexity and the reconstruction of the world. *Journal of Industrial Ecology, 13*(2), 168–183.

Allenby, B. R. (2011). Geoengineering: A critique. *Proceedings of the 2001 IEEE Annual Symposium on Sustainable Systems and Technology*, Chicago, IL, May 16–18, 2011.

Allenby, B. R. (2012). *The theory and practice of sustainable engineering.* Upper Saddle River, NJ: Pearson Prentice-Hall.

Allenby, B. R., & Sarewitz, D. (2011). *The Techno-Human Condition.* Cambridge: MIT Press.

Beattie, A. (2009). *False economy: A surprising economic history of the world.* New York: Riverhead Books.

Bijker, W. E., Hughes, T. P., & Pinch, T. (Eds.). (1997). *The social construction of technological systems.* Cambridge: MIT Press.

Bruchey, S. W. (Ed.) (1980). *Small business in American life.* New York: Columbia University Press.

Castells, M. (2000). *The rise of the network society* (2nd ed.). Oxford: Blackwell Publishers.

Clark, A. (2003). *Natural-born cyborgs.* Oxford: Oxford University Press.

Cronon, W. (1991). *Nature's metropolis: Chicago and the Great West.* New York: W. W. Norton and Company.

Elvin, M. (2004). *The retreat of the elephants: An environmental history of China.* New Haven: Yale University Press.

Freeman, C., & Louca, F. (2001). *As time goes by: From the industrial revolutions to the information revolution.* Oxford: Oxford University Press.

Garreau, J. (2004). *Radical evolution*. New York: Doubleday.

Graedel, T. E., & Allenby, B. R. (1995). *Industrial ecology* (1st ed.). Upper Saddle River, NJ: Pearson Prentice-Hall.

Graedel, T. E., & Allenby, B. R. (2010). *Industrial ecology and sustainable engineering.* Upper Saddle River, NJ: Pearson Prentice-Hall.

Grubler, A. (1998). *Technology and global change.* Cambridge: Cambridge University Press.

Heidegger, M. (original essays 1952–1962, collected in 1977). *The question concerning technology and other essays.* (W. Lovitt, Trans.). New York: Harper Torchbooks.

Hendrickson, C. T., Lave, L. B., & Matthews, H. S. (2006). *Environmental life cycle assessment of goods and services: An input output approach.* Washington, DC: RFF Books.

Kennedy, P. (1989). *The rise and fall of the great powers.* New York: Vintage Books.

Leopold, A. (1949). *A sand county almanac.* Oxford: Oxford University Press.

Marx, L. (1964). *The machine in the garden: Technology and the pastoral ideal in America.* Oxford: Oxford University Press.

Nature editorial. (2003). Welcome to the anthropocene. *Nature, 424,* 709.

Nye, D. E. (2003). *America as second creation: Technology and narratives of new beginning.* Cambridge: MIT Press.

Nye, D. E. (1994). *American technological sublime.* Cambridge: MIT Press.

Rittel, H., & Webber, M. (1973). Dilemmas in a general theory of planning. *Policy Sciences, 4,* 155–169.

Rosenberg, N., & Birdzell, L. E., Jr. (1986). *How the West grew rich: The economic transformation of the industrial world.* New York: Basic Books.

Schivelbusch, W. (1977). *The railway journey: The industrialization of time and space in the 19th century.* Berkeley: University of California Press.

Stock, G. (2003). *Redesigning humans.* New York: Mariner Books.

# Chapter 4

# Technology Mandates and Socio-Behavioral Life Cycle Assessment

**Gary E. Marchant**
*Sandra Day O'Connor College of Law, Arizona State University, Tempe, Arizona, USA*
gary.marchant@asu.edu

## Introduction

New technologies are often necessary to achieve evolving societal goals, such as reducing environmental harms, improving industrial or economic efficiency, or enhancing communication, educational, or entertainment capabilities. While market forces will usually create incentives to develop technologies that can meet these societal demands, there sometimes can be a gap between what the market is capable of producing and the broader societal interest. These market failures and barriers, whether real or perceived, often drive legislatures and regulatory agencies to enact policies to accelerate the development, adoption, and use of beneficial new technologies that they conclude are needed to achieve societal goals.

In the past government has used a number of tools to try to induce beneficial technology change, ranging from generic incentives for

*Emerging Technologies: Socio-Behavioral Life Cycle Approaches*
Edited by Nora Savage, Michael Gorman, and Anita Street

ISBN 978-981-4411-00-4 (Hardcover), 978-981-4411-01-1 (eBook)
www.panstanford.com

innovation such as research funding, a research and development tax credit, and intellectual property protection, to more targeted measures such as government procurement policies for specific products, consumer tax credits, and regulatory requirements (Marchant, 2009). Perhaps the most aggressive example of government technology inducement efforts until at least recently was the technology-forcing regulation of motor vehicle emissions, which provides vehicle manufacturers a specified lead time to develop new technologies needed to meet performance levels unattainable at the time of adoption (NRC, 2006).

Over the past decade, governments have increasingly adopted an extraordinary new regulatory tool for attempting to induce beneficial technological change. This tool involves the direct mandating of specific technologies, and recent U.S. examples include technology mandates (whether express or *de facto*) for electric vehicles (EVs), digital TV (DTV), and compact fluorescent lightbulbs (CFLs). This new tool of mandating specific technologies represents a radical departure from the evolved conventional wisdom in modern U.S. regulatory theory and practice that regulators should focus on performance rather than design standards, leaving it to the creativity of the market to select the best technologies and approaches for achieving the performance goals set by the government (Jaffe *et al.*, 2002; Stewart, 1981).

Yet, for a variety of reasons, governments have recently felt compelled to mandate specific technologies that are viewed as essential for meeting important societal goals. Such measures are risky because they require the governments to "pick winners" among competing technologies as well as the timeline in which that technology can be brought to fruition, both of which combined require an accurate assessment of the full spectrum of benefits, costs, and consequences of different technology choices. Such attempts to "pick winners" are characterized by uncertainty and complexity (Nelson & Langlois, 1983). Thus, any government attempt to force specific technologies should only be undertaken, if at all, after a careful and detailed analysis of the technological, business, economic, legal, social, and psychological dimensions of the entire technology life cycle.

In this chapter, after exploring in more detail the technology mandate tool generically, I critically assess three recent governmental technology mandates—for EVs, DTV, and CFL bulbs. While these technology mandates have achieved some progress in achieving their desired goals, they have also encountered rough turbulence and unanticipated problems because of the government's failure to consider the full range of impacts and consequences of the technology mandate. The lesson from this analysis is that if government chooses to mandate additional technologies in the future, it must do a better job in anticipating and addressing the full range of environmental, economic, and social implications across the technology's life cycle, which would involve undertaking an assessment of these implications, a process I refer to as socio-behavioral life cycle assessment.

## Technology Mandates as a Regulatory Tool

The recent trend in regulatory practice is toward greater market flexibility and less government micromanagement to achieve regulatory objectives. Thus, performance standards that specify the overall level of performance a regulated entity must achieve, without dictating the means by which such compliance must be obtained, are preferred over design standards that stipulate a particular technology or method a company must use to achieve the applicable performance objective. Performance standards give industry more flexibility to innovate to find cheaper and better means to achieve compliance, often in the form of new technologies, that government experts may not have been able to anticipate at the time the regulation was enacted. Market-based approaches such as cap-and-trade and marketable permit systems shift even more discretion to private actors, with government in effect specifying overall market performance rather than company-specific performance, again to maximize the role that industry flexibility and creativity can play in achieving societal goals at the lowest cost (Jaffe *et al.*, 2002; Magat 1979).

Technology mandates therefore run counter to the current trend away from government control and inflexibility. Governments nevertheless have recently been resorting to technology mandates in limited circumstances to require the development of a targeted technology (although as discussed below a *de facto* mandate is sometimes framed as a performance standard). Not surprisingly, such technology mandates tend to be highly controversial, often portrayed as an unduly intrusive intervention of government into the marketplace.

Nevertheless, governments have chosen such an extraordinary regulatory instrument in certain situations that involve a combination of two influences. First, technology mandates are favored for technologies that seem particularly beneficial and attractive in achieving an important governmental objective and societal goal. Second, technology mandates are considered where there appears to be market failures or barriers that seem to preclude market-based development of the favored technologies. An example is complementary technology problems that create a "chicken-and-egg" dilemma where several different components made by companies in different industries are required for a technology to succeed, and no one company wants to "go first" and take the risk that companies making the complementary technologies will follow suit.

Nevertheless, technology mandates are risky and ripe with potential pitfalls and failure modes. A study of historical attempts by governments to "pick winners" in technology competition concluded that "[h]ere the historical record seems, for a change, unequivocal. Unequivocally negative" (Nelson & Langolois, 1983, p. 817). Accordingly, government attempts to mandate specific technologies need to be undertaken cautiously, if at all, with a full and thorough attempt to anticipate potential consequences and problems in advance. Unfortunately, the three examples assessed here, as have so many past attempts to force specific technologies, fail to provide the type of holistic life cycle assessment that is essential for such programs to succeed.

## Example 1: California's Electric Vehicle Mandate

California's largely unsuccessful attempt to mandate substantial EV sales by 1998 is the first example assessed here of a governmental technology mandate. California is given unique authority, compared to every other state, under the Clean Air Act (CAA) to set its own vehicle emission standards, subject to approval by the U.S. Environmental Protection Agency (EPA), due to its "pioneering" role in regulating vehicle emissions and the "compelling and extraordinary" air pollution problems it faces (S. Rep. No. 403, 1967, at 33). Under this authority provided in CAA section 209(b), California had adopted a series of progressively more stringent sets of emission standards throughout the 1970s and 1980s that ended up being precursors for subsequent federal standards.

In 1990, California adopted its most stringent vehicle emission program ever, known as the low-emission vehicle (LEV) program, that required vehicle manufacturers to sharply reduce emissions over a multiyear period, with compliance being measured by the fleet-wide average emission rate for each manufacturer. This approach of the LEV program provided a broad performance standard that gave vehicle manufacturers broad flexibility to craft their own compliance strategies. Perhaps due in large part to the flexibility provided by the program, the LEV program overall has been an enormous success, achieving reductions in emissions and advances in emission technology beyond those even anticipated at the time the aggressive LEV program was adopted in 1990 (NRC, 2006). As the lead California regulator commented 13 years after the adoption of the program, "we've seen the near impossible accomplished with gasoline vehicles: zero evaporative emissions, exceedingly clean exhaust—cleaner, in some cases, than the outside air entering the cabin for ventilation purposes, and emission control systems that are twice as durable than their conventional forebearers, forecasted to last an astonishing 150,000 miles" (CARB Chairman Dr. Alan Lloyd, quoted in CARB, 2003).

The LEV program had one requirement that was not as successful, however, which was a requirement superimposed on the overall fleet average requirement that required major vehicle manufacturers to deliver for sale in California a specified percentage of "zero-emission vehicles" (ZEVs) beginning at 2 percent of the manufacturer's total California vehicle sales in model year 1998, and increasing to 5 percent in 2001 and 10 percent in 2003. Although the requirement was phrased in technology-neutral language, the reality was that EVs were the only possible technology that could possible achieve zero emissions as defined by the regulation, within the timeline provided (Bedsworth & Taylor, 2007). As the California regulators noted in adopting the ZEV mandate, "[a]t present, only battery-powered electric vehicles are candidates to be ZEVs, although it is conceivable that other technology (e.g., fuel cells) could be developed in the future" (CARB, 1990, p. 5). The ZEV mandate was therefore a *de facto* mandate to produce and sell EVs and represented a significant and controversial departure from the flexibility otherwise provided in the LEV program.

The California Air Resources Board (CARB), the California state agency responsible for adopting and implementing air pollution standards, decided to mandate EVS for several compelling reasons. First, CARB determined that the only way it could possibly achieve the extremely challenging goal of ensuring healthy air levels in the Los Angeles basin was to reduce emissions from consumer driving to essentially zero and thus saw the ZEV mandate as an essential first step to that goal. CARB noted that EVs not only produced no exhaust emissions, but they also produced no evaporative or refueling emissions, which were other significant sources of air pollution from internal combustion engines (CARB, 1991, p. 48). Moreover, degradation of emission control performance in internal combustion vehicles as the vehicles aged had become a major problem in efforts to limit air pollution, a problem that the ZEV mandate could help address since EVs produce zero emissions throughout their lifetime.

CARB had determined that market-based or other more flexible approaches for encouraging EVs were insufficient because battery manufacturers would not have the assurance of an adequate demand for EV batteries they needed to build commercial production capacity, and utilities and municipalities would not have sufficient assurance to install the necessary EV recharging capacity. By mandating the production of sufficient numbers of EVs, CARB hoped to solve these complementary technology roadblocks. Consumer demand for EVs had not yet developed—and may not have ever developed if EVs remained costly and a novelty—but by mandating large number of EVs to be produced and offered for sale CARB hoped to use "demand stimulus" to overcome the "latent" public demand by creating an economy of scale to drive down prices and to put sufficient vehicles into service to overcome any novelty concerns. Finally, California hoped that the ZEV mandate would provide an important economic boost for the start-up of clean-production companies in California and nationwide. Thus, CARB hoped that the ZEV mandate would "contribute to national and state energy diversity and security" efforts and help "to revitalize California's economy through job creation and growth in an emerging industry" (CARB, 1994a).

California's adoption of the ZEV mandate in 1990, scheduled to first apply in model year 1998, opened the door to other states to adopt the same requirement. Under section 177 of the CAA, other states may adopt vehicle emission standards "identical" to California's rather than applying the less stringent federal standards.

After California adopted its LEV program in 1990, several other states, most notably Massachusetts and New York, adopted the same program, including the ZEV mandate.

There were doubts and controversy about the feasibility and wisdom of California's ZEV mandate when it was adopted in 1990, and those concerns continued to grow in the following years. When it first adopted its ZEV mandate, CARB established a biennial review process to assess the industry's progress in developing ZEVs. In preparation for the 1996 review, CARB created an independent panel of battery experts to evaluate the readiness of battery technology for EVs. In its final report issued in December 1995, the panel determined that most vehicle manufacturers would try to meet the ZEV mandate in the 1998 model year using EVs with lead–acid batteries (Battery Technical Advisory Panel, 1995). The panel found that EVs powered by lead–acid batteries would have range and cost limitations that would restrict them to "small market niches" (*ibid.*, p. IV-1). The panel concluded that commercially successful EVs would require "advanced" batteries, but those batteries, using formulations other than lead–acid, were still at the prototype stage of development (p. iii–iv). In the best-case scenario, the panel determined that EVs with "commercial-production advanced batteries" would not be available until model year 2000 or 2001 at the earliest (*ibid.*, p. IV-5).

In response to the independent experts' report, CARB concluded that going forward with the ZEV mandates for the period prior to model year 2003 would be counter-productive: "[T]here is a substantial risk that due to various factors, including vehicle performance, e.g., range, and expected initial vehicle price, consumer demand may not support a market meeting the current regulatory requirements for model years 1998 through 2002, or alternatively that ZEVs introduced during that period may not meet consumer expectations and that the adverse effect of this situation may impact the viability of the market when advanced battery technologies are available" (CARB, 1996a). CARB was concerned that large-scale introduction of lead–acid ZEVs could "poison" the long-term prospects for ZEVs (CARB Final Statement 1996b, p. 7), and ultimately determined that it would not be "prudent to rely upon a large scale introduction of lead-acid battery EVs to launch the consumer ZEV market" (CARB, 1996c, p. 22). Based on those determinations, CARB amended its regulations in 1996, while vehicle manufacturers were in the process of finalizing vehicle production plans for model year 1998, to repeal the ZEV mandate for model years 1998–2002.

The remaining 2003 ZEV mandate was progressively weakened and eventually eviscerated in subsequent annual reviews. In the 1998 biennial review, CARB relaxed the 2003 mandate by permitting manufacturers to meet the majority of their 2003 ZEV target with "partial ZEVs" (PZEVs) ) (in effect, advanced hybrid vehicles), which were a new category of vehicles achieving near-zero emissions. In 2000, CARB appointed another panel of independent experts to assess battery performance and availability, which concluded that advanced technology batteries with adequate cost and performance characteristics would not be available to meet the 2003 mandate (Year 2000 Battery Technology Advisory Panel, 2000). After much indecision and several flip-flops on how to respond to the battery experts' report, CARB decided again at the last minute in January 2001 to further weaken the 2003 mandate. Soon thereafter, manufacturers filed a successful lawsuit contending that the ZEV program was preempted by federal law, which resulted in temporarily enjoining the program. CARB settled the lawsuit by essentially eliminating the original ZEV mandate, allowing manufacturers to meet their 2003 obligation by producing a small number (250 for the total industry) of demonstration fuel cell vehicles by 2008, followed by a ramp up of fuel cell vehicles development in subsequent years.

This final nail in the coffin of the original ZEV mandate effectively replaced the original mandate for EVs with a requirement for hybrid vehicle sales in the short term and demonstration projects for fuel cell vehicles in the long term. In throwing in the towel on EVs, CARB concluded that "even at substantially increased production levels full function EVs would not be cost competitive with conventional vehicles, and that there does not appear to be a viable path that will result in commercialization for general markets" (CARB, 2004, p. 51). "Under these circumstances," CARB conceded, "it is not surprising that the major automobile manufacturers express an unwillingness to develop or produce additional full-function battery EVs" (CARB, 2004, p. 17). In all, the ZEV mandate was relaxed six times, progressively diluting the program into a pale imitation of its original grandiose goals and requirements, although not without imposing substantial compliance costs as companies invested enormous resources struggling to try to meet the infeasible ZEV mandate (NRC, 2006).

Ironically, shortly after CARB abandoned its focus on EVs, vehicle manufacturers started showing renewed interest in EVs. Several vehicle manufacturers announced plans to commercialize EVs, including the Nissan Leaf, the GM Volt, and the Toyota plug-in Prius. But, as CARB itself noted, the driver of this renewed focus on EVs was not CARB's ZEV mandate, but rather that "long-term commercial success and corporate environmental stewardship" had become the primary motivations for manufacturers (CARB, 2009, pp. 14–15). To be sure, the 1990 California mandate no doubt had some incremental beneficial impact in encouraging the development of EV batteries and other technologies that now seem to be coming to fruition (Bedsworth & Taylor, 2007). A study commissioned by CARB found that the ZEV mandate did provide significant secondary benefits in the development and applications of battery and other technologies that benefited both hybrids and EVs (Burke *et al.*, 2000).

Yet, notwithstanding these indirect benefits, the California ZEV mandate was a monumental failure. Hundreds of millions if not billions of dollars were wasted by vehicle manufacturers in frantic programs to attempt to rush EVs to market prematurely in order to comply with the ZEV mandate (NRC, 2006). While some such vehicles did see the light of the market, such as the GM EV1 and the Honda EV Plus, they encountered very low consumer demand, primarily limited to niche markets of EV enthusiasts, and were soon cancelled. As Alan Lloyd, the Chairman of CARB at the time the original ZEV mandate was finally repealed, ruefully conceded, "We have put a lot of faith in battery electric vehicles to meet the [zero-emission vehicle] mandate but, in spite of significant efforts, batteries have inherent limitations" (Polakovic, 2002).

The failure of the California ZEV mandate can be attributed to a number of factors. To begin with, CARB faced a difficult task in picking both the winning technology and a realistic timeline. Several clean vehicle technologies were on the horizon, including hybrids, alternative-fueled (e.g., natural gas, propane, or ethanol) vehicles, battery EVs, and fuel cell vehicles. CARB decided to place all its bets on EVs—rejecting manufacturer suggestions in the early 1990s to give partial credits for hybrid electric vehicles (HEVs) (NRC, 2006). CARB stated:

> The adopted regulations reflect a conservative stance toward special provisions for HEVs because, unlike ZEVs, HEV emissions are expected to deteriorate in-use due to the existence of the auxiliary engine. Also,

> as batteries deteriorate in HEVs, use of the auxiliary engine would probably increase. Incentives to produce HEVs, such as allowing HEV credits to substitute for the ZEV mandate, will delay battery development for better performing ZEVs (and, therefore, HEVs) (CARB, 1991, p. 51).

Moreover, the compliance schedule adopted by CARB was not feasible for fuel cell vehicles and thus effectively excluded them from its *de facto* mandate for battery EVs. Unfortunately, the aggressive timeline CARB selected turned out, at least in retrospect, to be unrealistic.

The difficulty of picking winning technologies and their development timeline make technology mandates risky, but if CARB was nevertheless determined to attempt such a mandate, it could have done a better job trying to anticipate and consider the full range of technological, economic, business, environmental, consumer, and social impacts and uncertainties about the technology it sought to advance. For example, one factor contributing to the program's failure was the unrealistically optimistic estimates that CARB held about EV costs, feasibility, and timeline. When it first adopted its ZEV mandate in 1990, CARB claimed that "by 2000, electric vehicles would be comparable in cost to conventional vehicles except for the additional cost of batteries," which CARB estimated at the time would cost only about $1,350 per EV (CARB, 1990, p. 63), a number that was far lower than independent estimates at the time. CARB likewise dismissed in 1990 concerns that EVs may not provide the performance most vehicle buyers expected and thus may not be commercially viable, stating that "there is little concern of the commercial viability and public acceptance of electric vehicles" (CARB, 1991, p. 48). Just four years later, in its 1994 review of the ZEV mandate, CARB had increased its estimate of the incremental additional costs of an EV from its initial estimate of $1350 per vehicle to a new estimate of $5000 (low scenario) to $10,000 (high scenario) (CARB, 1994b, p. 65). By the 2000 program review, CARB had increased its estimate of the incremental cost of a freeway-capable ZEV to at least $17,000 more than a conventional vehicle, more than an order of magnitude higher than the estimate CARB had provided at the time it adopted its ZEV mandate in 1990 (CARB, 2000, p. 127).

This increasing cost estimate was due in part to CARB's erroneous assumption in 1990 that current (i.e., lead–acid) battery technologies would be adequate for a commercially viable EV in the initial years of the ZEV mandate, as well as an underestimate of the challenges and timeline for the availability of advanced technology batteries. By 2010, over a decade after the CARB ZEV mandate was initially scheduled to take effect, the cost and performance of EV batteries were still not predicted to become competitive with the internal combustion engine for several more years if not a decade or more (MIT, 2010). More broadly, CARB's flawed cost and performance estimates suggest a failure to seriously consider the full range of opinions and uncertainties about the future of EVs, as the agency adopted uncritically the most optimistic opinions and estimates that aligned with its own preferences while dismissing without serious consideration more critical estimates from industry and other experts.

CARB also failed to consider the perspectives of key stakeholders. EVs could only be a success and provide the air quality benefits CARB was seeking if consumers felt comfortable buying and driving EVs and would be willing to switch to EVs in large numbers. Other than small niche markets, most consumers had several concerns about EVs, including dislike of their look and design, worries about the reliability of unfamiliar technologies, dissatisfaction with overall power and performance, anxiety about driving range, and concern about the time needed to recharge battery packs (J.D. Power, 2010). As a recent MIT conference report succinctly stated the issue: "for EVs to successfully penetrate US markets, consumer acceptance is essential. Without this, the answer to the question 'Who killed the electric car?' will likely not involve a sinister conspiracy but will instead simply be 'the American consumer'" (MIT, 2010, p. 15).

Yet, the artificial timeline that CARB imposed exacerbated rather than helped to address these consumer concerns by forcing the premature production of EVs that cost more, performed worse, and had many more uncertainties than a conventional gasoline-fueled vehicle. In addition, EVs were not appropriate for many consumers, including those who needed their vehicle for trips greater than the approximately 100 mile maximum range of an EV (under ideal conditions), or those who did not park their car in a garage or other location with a nearby recharging capability. Indeed, studies available at the time that CARB adopted its ZEV mandate indicated that vehicle range was the most critical factor for consumers in deciding whether to purchase EVs (Calfee, 1985).

Yet, CARB simply ignored concerns about the limited range of EVs and the widespread "range anxiety" that made EVs unattractive to most consumers. Not surprisingly, EV purchases were limited to the small niche of EV enthusiasts, and the initial EVs produced under the ZEV mandate (before it was essentially repealed) sold (or were leased) in the hundreds of units rather than the tens or even hundreds of thousands needed to make a major impact on air quality. CARB seemed to assume that if it could force the supply of EVs, the demand for EVs would follow, a premise that turned out to be faulty at least without a lot more effort to build consumer demand. EVs, especially with the performance limits in terms of range and other attributes that existed in the late 1990s and first decade of the 2000s, were not going to sell themselves in any significant numbers.

CARB also failed to consider the incentives and interests of the companies it depended on for the success of the ZEV mandate. By mandating that every major vehicle manufacturer produce the same percentage of EVs in the same time period, CARB unwittingly undermined any competitive advantage that a vehicle manufacturer might gain from seeking to be an industry leader in the EV field. Thus, General Motors (GM) was actively developing and planning to commercialize its own EV, originally named the Impact, before CARB decided to mandate EVs. By imposing a rigid and what appeared to many (including those in industry) an infeasible EV mandate, CARB turned GM from a proponent of the vehicle it was developing on its own to an EV (or at least an EV mandate) critic. Thus, one of the strongest proponents of EVs became one of their strongest opponents, and CARB turned a potential ally on the commercialization of EVs into a well-funded and determined opponent.* CARB also failed to understand the dynamics and incentives of battery manufacturers, simply assuming that if it mandated the vehicles then battery manufacturers would quickly and automatically fall in line and start mass producing the batteries needed by the vehicle manufacturers. The reality was much more complicated given the uncertainties about battery types and performance needed for commercially successful EVs, and given the fluid situation and changing perspectives

*The author represented GM as an attorney on EV issues in this time period and observed first-hand the counter-productive impact of the ZEV mandate inside the company. Although GM tried to distinguish its opposition to the ZEV mandate from its support and interest in EVs, those two positions were hard to reconcile in both internal and external contexts.

about the best battery technologies, battery manufacturers had no assurance that vehicle manufacturers would purchase any particular battery technology the battery manufacturer would have to commit to in building mass production capacity.

CARB also failed to address the full life cycle environmental impacts of EVs. When it adopted the ZEV mandate, CARB emphasized the benefits of EVs in producing zero emissions throughout their life cycle, including from the tailpipe (especially as the vehicle aged), from evaporative emissions, and from refueling. Notwithstanding these clear environmental benefits, there remains the issue of environmental impacts from the electricity generation used to charge the EVs (Gruenspecht, 2001). This has proven to be a controversial and contested issue, perhaps due in part to CARB's failure to address this issue openly and fully at the time it adopted its ZEV mandate. While most analyses show that EVs in California would provide a clear environmental benefit given the relatively clean fuels used to generate electricity in California, that may not be the case for other States that also adopted a ZEV mandate identical to California's but which rely much more heavily on coal to generate electricity (Sperling & Gordon, 2008). In fact, a recent National Research Council report concluded that the adverse health and environmental impacts of EVs in 2030 under various assumption and scenarios would be consistently worse than for conventional vehicles (NRC, 2010).

CARB also did not address other potential environmental concerns related to EVs, such as the disposal and recycling of the large battery packs used by EVs. Almost all EV manufacturers are currently planning on using lithium batteries (MIT, 2010), yet few recycling companies are currently accepting lithium batteries from laptops and cell phones, raising concerns that lithium batteries from EVs may create disposal problems (Ramsay, 2010). A recent analysis of the life cycle impacts of EVs conducted for the Low Carbon Vehicle Partnership in the EU found that the production of EVs would produce substantially more $CO_2$ than the production of conventional vehicles due to the battery pack production, although the EVs produced lower whole-life overall carbon emissions (assuming it was able to complete its life span without a replacement battery pack) (Ricardo, 2011). Another factor overlooked by the ZEV mandate is that a shift to EVs would have a negative impact on water resources, primarily due to increased water cooling of thermoelectric power plants (King & Webber, 2008).

Finally, CARB did not consider other social and economic second-order impacts of the EVs it was mandating. One such impact is that because EVs do not need to purchase gasoline, they do not pay the highway tax built into the price of gasoline that is used to fund highway construction and maintenance in this country. As the highway fleet shifts toward electric and hybrid vehicles, this could present a serious budgetary problem. Already some jurisdictions (e.g., Oregon, Texas) are considering imposing a vehicles miles traveled (VMT) tax on EVs, using a GPS chip in the car to keep track of mileage (Hurst, 2011). The mandatory GPS chip would raise privacy and access concerns.

In his 2011 State of the Union speech, President Obama called for one million EVs to be on the road in America by 2015. Every major vehicle manufacturer has one or more EVs in commercial development, and so achievement of the one million vehicles by 2015 is not an impossible national goal, although there are many doubts about the feasibility of the timing. For example, a recent National Research Council report concluded that "a significant market penetration" of EVs prior to 2030 "is unlikely" (NRC, 2010, pp. 198–199). An October 2010 assessment by J.D. Power and Associates estimates that there will be only 100,000 battery-powered EVs sold in the United States in 2020 (J.D. Power, 2010). The cost of batteries, one of the chief impediments to wider commercialization of EVs, is not expected to significantly decrease for at least several years and may, in fact, increase due to shortages of many of the key metals required for battery production (Ramsay, 2010). Nevertheless, most experts believe EVs will be the vehicles of the future, it is a question of when not whether (Sperling & Gordon, 2008).

Yet, regardless of the projections about the future, what is no longer in doubt is that California's attempt to force more than one million EVs on the road in California alone by 2003 was plainly premature. Although pursuing a legitimate (and perhaps necessary) objective, CARB's failure to provide a realistic scenario and timeline for EVs was due to its failure to consider the full range of economic, business, consumer, social, and environmental impacts and challenges of EVs. A fuller and more robust analysis up front might have avoided many of the pitfalls and wasted resources and opportunities that are the legacy of the failed ZEV mandate.

## Example 2: Digital TV Mandate

The second example of a technology mandate was the Congressional requirement to transition television from analog to digital broadcast signals by 2009. After several delays, this transition was eventually accomplished, so the statute achieved its ultimate goal, but not without a lot of controversy, disruption, and inefficiencies that could have been avoided or at least reduced by a more thorough and deliberate anticipation of potential technological, economic, business, social, and consumer impacts and responses.

DTV offers a number of advantages over analog TV. It enables better quality picture and sound transmission, free of the "ghosting" and fuzziness that often afflicts analog broadcasts. It also permits interactive services such as viewer-controlled feedback or control, as well as other digital computer-like voice and data applications including emailing and software delivery. DTV also permits more exotic services such as "multicasting" that allows several different feeds to be transmitted simultaneously to a viewer (enabling the viewer, for example, to watch several basketball games on the screen at the same time). Perhaps most importantly to government, digital broadcasts utilize significantly less spectrum than analog signals, freeing up valuable spectrum that could be utilized for other important uses (and government revenues).

Yet, despite these important benefits, the government and many industry experts concluded that the market would not transition to the superior digital technology in a timely manner on its own. The major obstacle was the "chicken-and-egg" problem with three complementary and mandatory industry participants—the equipment manufacturers needed to produce DTV sets, the content providers needed to produce digital versions of TV shows, and the television stations needed to broadcast digital content. It was not in the interest of any one of these three distinct industry sectors to break the deadlock and "go first" and begin the transition to digital. For example, why should a content provider produce a digital version of a TV show if there were no TV stations capable of transmitting digital content and no consumers having DTVs capable of receiving such content? This chicken-and-egg problem among complementary technologies was exacerbated by the latent public demand for DTV—

since consumers had generally little or no experience or knowledge about the benefits of DTV, there was very little consumer push for the technology.

To break this logjam, the Federal Communication Commission (FCC) and Congress mandated that all broadcast TV channels switch to digital broadcasts by a specified date, which was delayed several times. The initial DTV transition plan was adopted by the FCC in 1998 after a decade-long proceeding to consider the future of advanced technology TV and required TV stations to start broadcasting a digital channel by May 2002, but also temporarily "loaning" each TV station a second frequency so that it could simulcast its programs on both an analog and digital channel for an interim period. The analog channel would be shut down at some point and the spectrum allocation returned to the government to be auctioned off at the completion of the transition. Congress passed a "Sense of Congress" resolution in 1997 specifying an original target date of December 31, 2006, but provided a "fail safe" mechanism that would delay the deadline if 85 percent of homes did not have equipment that could handle a digital signal. This safety valve was indeed triggered, with less than three percent of American households having TV sets capable of receiving digital signals by early 2005 (although many more could receive digital broadcasts through their cable or satellite systems), and accordingly the FCC delayed the implementation for two more years to December 31, 2008 (Hart, 2010).

The Digital Television Transition and Public Safety Act of 2005 extended the deadline by a few weeks to February 17, 2009 in order to avoid inconveniencing viewers during the holiday football bowl games and Super Bowl. By January 2009, just a few weeks before the scheduled cutoff of the analog signals, the DTV transition was in a state of "disarray" and "turmoil," according to the *Washington Post* (Hart, 2009c). Facing widespread opposition and lack of preparation, a divided Congress once again extended the deadline in the DTV Delay Act of 2009 to June 12, 2009. Stations were given the option of proceeding with the original February 2009 date, and approximately 40 percent of U.S. stations made the transition at that time (Boyce, 2009). When the final transition occurred in June 2009, an estimated 11–12 million homes (10 percent of U.S. households with TVs) were still at risk of losing TV reception and had not installed a converter box (Labaton, 2009; Stelter, 2009a,b).* Nevertheless, the reported complaints and disruption were relatively minor according to some accounts, described by one reporter as a "minor hiccup" (Stelter, 2009a). However, the FCC call center received almost one million calls per weeks around the time of the transition (Stelter, 2009b). Moreover, the people who were most likely to have lost their TV reception tended to be poor, non-English speaking, undereducated, and rural residents, and they may be the least knowledgeable and least capable of registering their complaints with the appropriate authorities (Labaton, 2009; Stelter, 2009b).

The mandated transition to DTV resulted in enormous confusion, resistance, and delays, notwithstanding its ultimate (delayed) success. Again, many of these problems were due to the government's failure to consider the full range and life cycle of potential impacts and responses. Perhaps most significantly, the government failed to anticipate and prepare to address the public response to having their analog TV cutoff. At the time the policy was being implemented, as many as 1 in 5 American households (Steinberg, 2007) received their TV broadcast through antennas and would lose all TV reception using those systems unless they purchased a converter. As one writer stated the challenge, "the 21 million households using a conventional set with rabbit ears or a rusty roof antenna—typically people who are poor, elderly, or living in rural America—will turn on their TVs and see ... nothing" (Antonoff, 2007, p. 71). One member of Congress described the risk as follows, "The government can do a lot of things, but you mess up people's televisions, and they're going to be very upset. It's going to be a real mess when it rolls out in the real world" (Hart, 2008a, p. D2, quoting Rep. Heather Wilson).

The FCC initially relied primarily on industry forces to educate consumers about the pending transition, and this effort was inadequate, resulting in widespread consumer ignorance or confusion about the transition (GAO, 2005). The chairman of the FCC announced a "major DTV consumer education initiative" called "DTV—Get It!" in October 2004, but the effectiveness of this program was uncertain and it was criticized for being uncoordinated (Hart,

*Approximately 3 million of these households rely on analog sources only and will lose all TV reception, while approximately 9 million households have one TV connected to a cable or satellite but other TV sets in the house that are not connected and will lose reception.

2009b).* In 2009, then Commerce Secretary Gary Locke criticized the government's education program for failing to provide adequate information to consumers (Labaton, 2009). For example, much of the educational campaign informed consumers they would need to buy a converter box to receive the new digital signals on their existing TV set, but the public was not told they likely needed a new antenna as well and would have to rescan their TVs to find the new channels (Hart, 2009b). One expert stated: "This transition is possibly one of the worst understood consumer education programs in modern times" (Taub, 2008, quoting Richard Doherty). Other jurisdictions such as Berlin and the United Kingdom had undertaken much more thorough public education campaigns in advance of their DTV transitions (GAO, 2005).

Facing the reality that many consumers either did not understand or were not willing to buy their own converter, Congress created a hastily crafted federally funded program to provide consumers a coupon for $40 (limited to two coupons per household) toward the purchase of a converter box. This subsidy program was limited by Congress to $1.5 billion (enough to fund 33.5 million coupons) (Dunbar, 2008). The agency official in charge of this program estimated that the demand would range from 10 to 26 million coupons (Dunbar, 2008), which substantially underestimated the demand for these coupons. By the end of 2008, 40 million coupons had been requested, but only 16 million had been redeemed (Hart, 2010). The government stopped providing coupons on January 5, 2009, as the authorized budget for issuing such coupons had been exhausted, with over 1 million consumers on the waiting list to get a coupon (Hart, 2009a). Congress authorized an additional $650 million for coupons in February 2009. By the time of the final transition, 59 million coupons had been issued and 31 million had been redeemed (Hart, 2010), far more than were anticipated by the program developers.

In January 2009, the incoming Obama Administration, supported by several consumer groups and members of Congress, proposed a further delay of the DTV mandate given the shortcomings of the coupon program and the fact that many Americans were still not prepared for the transition scheduled just one month from

---

*A poignant anecdote that perhaps summarizes the public education program is that at one point the government sponsored a NASCAR race car emblazoned with the logo "The Digital TV Transition," but the car promptly crashed in a race (Taub, 2008).

then. According to the letter sent to Congress by the cochair of the incoming administration's transition cochair, "with coupons unavailable, support and education insufficient, and the most vulnerable Americans, I urge you to consider a change in the legislatively-mandated analog cutoff date" (John Podesta, quoted in Hart, 2009a). Congress passed a 4 month extension of the deadline signed by President Obama on February 11, less than one week before the deadline was to take effect. In signing the legislation, President Obama stated that "Millions of Americans, including those in our most vulnerable communities, would have been left in the dark if the conversion had gone on as planned" (Hart & Whoriskey, 2009).

The government also failed to accurately anticipate how companies in the various industry sectors would respond to the DTV mandate. Most TV stations, the industry sector directly subject to the digital mandate, did not comply with the original deadline to start broadcasting a digital station (Labaton, 2002; Taub, 2002). The cost of the transition, particularly for smaller stations, along with the lack of any perceived benefit given the rarity of consumers with DTV sets, was responsible for the tardy compliance. The National Association of Broadcasters, the trade association of TV stations, actively lobbied for a delay in the implementation of DTV (Shiver, 2001).

TV equipment manufacturers resisted phasing out the production and sale of analog sets and, even after the government's adoption of its DTV transition plan, continued to sell 25 million analog TVs per year. The FCC finally had to step in, and in 2002 it required equipment manufacturers to stop marketing technologically obsolete analog sets by 2007. The manufacturers sued the FCC to try to stop this part of the mandate but were rejected by the court (*CEA v. FCC*, 2003). Cable and satellite providers were not subject to the digital mandate, even though they provided programming to 80 percent of U.S. homes (Taub, 2002). These companies objected to pressure that required them to carry both analog and digital channels without necessarily earning more revenue and were finally forced to carry both analog and digital signals by the FCC "must carry" rules that were adopted relatively late in the game (Hart, 2010).

The government also failed to anticipate and address performance problems with DTV, which were predictable at the time the policy was adopted. Digital broadcast suffer from a limitation known as the "digital cliff"—if there is any interference with a signal, the screen goes blank, instead of becoming fuzzy like analog TV. Also, digital signals do not travel as far as analog signals (Hart, 2008b). Thus, many consumers found that there new digital systems failed to obtain proper reception either because they lived outside the smaller coverage area of a station or the signal was partially obstructed by buildings, a tree, or other objects, or even bad weather (Hart, 2008b; Hart, 2009b). The DTV antennas also require more precise positioning than their analog predecessors (Hart, 2008b). The federal government did not make any serious effort to educate consumers about these reception gaps, creating even more consumer frustration with the transition (Hart, 2008b).

Finally, the U.S. government failed to consider up front and try to limit the environmental impacts of its DTV mandate. Many consumers were motivated to buy new DTVs (as opposed to the other option of buying a converter for their existing TV), and consequently millions of now obsolete analog TVs were discarded into the trash. While proper recycling of TVs is possible, there was little or no organized effort to educate consumers about this during the DTV transition. As one environmentalist reported, "we are now seeing a huge anticipated spike in the amount of electronic waste, really a tsunami of electronic waste coming through because of this digital transition" (Palm, 2009). The federal government completely ignored this issue until the EPA launched a campaign entitled "The Digital Television Transition: Don't Kick Your TV to the Curb—eCycle Instead," but this effort was not launched until June 11, 2009, just one day before the DTV transition was completed (EPA, 2009).

Overall, the U.S. digital transition was heavily criticized for how it was handled by the government (Hart, 2010; Taub, 2008). Critics described the program as a "disastrous failure of public policy" (Hazlett, 2001), "a train wreck" (Hazlett, 2001), "fiendishly complex" (Farhi, 2009), and "more like Hurricane Katrina" (Hart, 2009a). There was confusion, modifications, and uncertainty throughout the ten-plus year transition process. As the FCC's own key official responsible for implementing the transition commented in the middle of the process: "We're midstream in a boat that some say has leaks. We're patching the leaks and trying to get to shore" (Taub, 2002, quoting Kenneth Ferree, Chief of FCC Media Bureau). A *Washington Post* writer commented: "The transition from analog to digital television may have inflicted more confusion on the American public than any other electronic upgrade in history. No other technological switch ... seems to have puzzled, frightened and in some cases angered so many people" (Pegoraro, 2009, p. D3). The repeated delays in the implementation of the transition, made necessary by the poor planning and anticipation of problems, not only sowed confusion amongst consumers but also imposed major problems and costs on the companies subject to the requirements and other affected parties (Hart, 2009c).

One key problem throughout was the lack of a focused, coherent, and consistent government policy. The GAO issued a scathing report criticizing the federal government for lacking a comprehensive plan to implement the DTV transition that would provide for a more coherent rollout and facilitate better planning and risk management (GAO, 2007). Indeed, the U.S. digital transition was so long, convoluted, and confusing that it had the effect of freezing innovation by affected industries, undermining much of the promised benefits of the DTV conversion (Farhi, 2009). As one commentator wryly noted, while digital technology was supposed to provide many exciting, innovative, and interactive benefits, the reality was quite different: "So after enduring loads of hassle—coupons! Deadlines! Converter boxes!—the net impact of the digital conversion will be a few more channels and the chance to see Mr. T again" (Farhi, 2009). The one consolation, according to the *Washington Post*, is that "after all the angst the digital transition has inflicted on home viewers and policy makers alike, it will be a long time before anybody suggests another mandatory upgrade over the public airwaves" (Pegoraro, 2009).

## Example 3: CFL Bulb Mandate

The third and final example of a technology "mandate" is the transition to more efficient lightbulbs. In the United States, and virtually every other industrialized country that have adopted similar measures (Waide, 2010), the transition to more efficient lighting has not involved any formal technology mandates or bans, but rather a performance standard with a phased implementation schedule that will have the effect of gradually eliminating traditional incandescent lightbulbs from the market (Logan, 2008). Specifically, the Energy Independence and Security Act (EISA) of 2007 established minimum efficiency and lifetime requirements that are phased in from January 1, 2012 through January 1, 2014. The practical effect of these increasingly stringent performance requirements will be to prohibit the 100 W incandescent bulb in 2012, the 75 W incandescent bulb in 2013, and the 60 W incandescent bulb in 2014 for general service applications, with some exceptions for special purpose and decorative products. Notwithstanding the deliberate performance-based phrasing of the legislation, its practical effect will be to ban the traditional incandescent lightbulb, and given the limited compliant replacement lighting technologies currently available, to *de facto* mandate the replacement technologies. Not surprisingly, therefore, the public and political debate about the legislation frequently uses the terms "ban" and "mandate," even though technically these terms are not accurate.

There is widespread expert agreement that the traditional incandescent lightbulb is a wasteful, obsolete technology and that the replacement with a more efficient technology holds great potential for substantial savings in cost, energy, and pollution prevention. The incandescent lightbulb is notoriously inefficient, converting less than five percent of the energy it uses into light while the remainder is emitted as heat (Azevedo *et al.*, 2009). Since lighting consumes more than 20 percent of all electricity generated in the United States, mostly by inefficient incandescent bulbs, transitioning to a more efficient lighting technology holds enormous potential and is seen as a "low hanging fruit" for saving energy and preventing pollution (including carbon dioxide) emissions (Azevedo *et al.*, 2009). The technology of incandescent bulbs is a mature technology, and very little technological progress has been made over the past 100 years in improving its performance, so it needs to be replaced rather than improved. The key issue is the availability and timing of suitable alternative technologies that are more efficient but which also satisfy consumer demands.

The largest anticipated replacement for the incandescent bulb and beneficiary of the legislation at least in the short- to medium-term was expected to be the CFL, seen by many as "the natural alternative" to traditional bulbs (Di Maria *et al.*, 2010; Scelfo, 2008). The CFL is more energy efficient than the incandescent bulb, and also lasts much longer. CFLs have an efficiency of approximately 13 percent, which means they are approximately five times more efficient than incandescent bulbs (Azevedo *et al.*, 2009). The Congressional Research Service estimates that a typical 100 W incandescent bulb used $18.30 in energy per year compared to only $4.9 for an equivalent CFL bulb, and the CFL bulb would last for 10,000 hours while the incandescent bulb only lasts 1000 hours (Logan, 2008). The CFL offers great potential to save energy, money, and environmental externalities such as $CO_2$ emissions from generating the electricity needed to operate the light.

While fluorescent lights have been used since the 1930s, they were not generally applicable to residential applications until the advent of the compact fluorescent bulb, first discovered in 1976 but made commercially available only in the 1990s as a result of technological advances such as the ability to cost-effectively manufacture bulbs consisting of tightly coiled gas-filled fluorescent tubes (Logan, 2008; Sandahl *et al.*, 2006). The CFL did not make major inroads into the residential market until the early 2000s when utilities and other entities began major programs to encourage CFL adoption. The market share of CFLs rose to a peak of 20% in 2007 with approximately 300 million CFLs sold in the United States, although their market share actually shrank to approximately 15% in 2008 and 2009 (Swope, 2010). This market penetration of CFLs has resulted in an expected decline in overall replacement shipments, as the CFL's longer life has reduced the frequency in which lightbulbs must be replaced (APT, 2010).

Yet, market public acceptance of CFLs has been slower than anticipated (LRC, 2003; Di Maria *et al.*, 2010), and major impediments are clouding the future of CFLs, factors barely considered, if at all, by Congress in adopting its phaseout of the incandescent bulb in favor of CFLs as the presumptive replacement.

Perhaps the biggest impediment to wider adoption of the CFL is the problems encountered with CFL performance and consumer satisfaction. These issues include (i) CFLs have a slower ramp-up to full luminosity compared to the standard incandescent bulb; (ii) most CFLs are not dimmable; and perhaps most significantly (iii) some consumers perceive the quality of light from CFLs as inferior to traditional lighting sources, with frequent complaints that the light is "too dim," "harsh and unflattering," "too blue," or otherwise "not right" (APT, 2010; Logan, 2008; LRC, 2003; Rice, 2011; Scelfo, 2008). For example, one focus group of consumers found that perceived color variations in CFLs was a major barrier to consumer acceptance (LRC, 2003). The study also found that consumers have a negative connotation with the word "fluorescent," likely a residue of the "unfriendly" fluorescent tube lighting used in many commercial establishments (LRC, 2003). Another recent survey of consumers found that the brightness of the light the bulb produces was the most important factor for consumers in evaluating lightbulbs, a finding that suggests some vulnerability for CFLs (KRC, 2010).

In addition, some CFLs and, in particular, many of the cheaper bulbs that were included in various utility and other incentive programs were of poor quality and did not meet expectations in terms of quality of light and life span. For example, some cheaper CFL bulbs had to be recalled because they presented a fire risk (e.g., CPSC, 2010). The extended life of the CFL is also sometimes not achieved in practice, further leading to consumer disenchantment and disappointment. For example, one analysis found that 2 to 13 percent (depending on brand) of CFLs failed early and that half of reflector CFLs used in recessed lighting had dimmed by at least 25% by halfway through their rated lifetime (Angelle, 2010). Another study performed for the California Public Utilities Commission found that the average useful life of a CFL in California was 6.3 years, considerably shorter than the projected useful life of 9.4 years (Smith, 2011). Moreover, frequently turning a CFL on and off (many times per day) will significantly reduce its life span (DOE, 2011).

Lightbulbs are perceived as low technology and low interest products by consumers, who want convenient, simple, and inexpensive replacement bulbs that do not present any "issues" or "problems" (APT, 2010). Consumers are also very sensitive to initial price, even if the longer life span of a CFL or other more efficient bulb will reduce replacement costs and cost less in the long term (APT, 2010; Di Maria *et al.*, 2010). Consumers apply a very high implicit discount rate—as high as 300 percent compared to the typical 2.5 to 10 percent used in most economic analyses—that deter consumer purchases of energy efficiency technologies that may cost more up front but save money over their lifetime due to lower energy and replacement costs (Azevedo *et al.*, 2009). This inflated consumer discount rate, sometimes referred to as the "energy paradox," is attributed to a number of factors, including lack of knowledge about cost savings, disbelief about lifetime savings, and lack of expertise in addressing the time value of money (Azevedo *et al.*, 2009; Jaffe and Stavins, 1994). Thus, it is not surprising that CFLs that cost more up front and raise various performance issues have encountered consumer resistance, especially if the net advantages of new lighting technology are not clearly communicated to consumers (Di Maria *et al.*, 2010). Yet, empirical studies indicate that up to 50 percent or more of consumers are unaware of the energy savings and environmental benefits of CFLs (Di Maria *et al.*, 2010; LRC, 2003).

Another important, unanticipated concern about CFLs is their potential environmental impact and, in particular, their mercury content. Each CFL contains a small amount of mercury (generally 3–5 mg per bulb) that forms the vapor inside the glass tube that fluoresces. EPA and others have pointed out that CFLs may still result in a net decrease in mercury releases into the environment because the mercury released from CFLs, especially if handled and disposed of properly, would be less than the amount of mercury emissions prevented from coal-fired power plants because of the increased energy efficiency of the CFLs (Energy Star, 2010). Nevertheless, if the mercury from tens of millions of CFLs accumulated in landfills or other inappropriate disposal routes, it could total a significant amount of mercury released and present a significant environmental and occupational exposure risk (Aucott *et al.*, 2003). As the *New York Times* editorialized: "[A]lthough one dot of mercury might not seem so bad, almost 300 million compact fluorescents were sold in the United States last year. That is already a lot of mercury to throw in the trash, and the amounts will grow ever larger in coming years. Businesses and government recyclers need to start working on more efficient ways to deal with that added mercury. ... Even when warned, public officials are never great at planning" (NY Times, 2008). As this editorial notes, governments seemed to not have anticipated or made any plans to address this predictable impact of the CFL-promoting legislation. While some states require recycling of fluorescent bulbs and some business have taken important steps to encourage free recycling, only 2 percent of residential users and approximately 30 percent of businesses currently recycle their CFLs (Bohan, 2011; Silveira & Chang, 2011).

Moreover, in addition to environmental release, there is a concern about exposure from individual bulbs that break in the home. The EPA recommends somewhat frightening-sounding special precautions in using and disposing of CFLs if they break (EPA, 2011), which may alarm some consumers. A European Union (EU) scientific advisory committee calculated that ambient room exposures to mercury are in the range of or exceed the occupational exposure limit (100 $\mu g/m^3$), but because that exposure limit is based on the safe level of *lifetime* exposure, the expert group concluded that adults would not be harmed by mercury exposures from a broken CFL bulb (SCHER,

2010). The expert group concluded that there was insufficient information to determine if the exposure levels would be safe for children (SCHER, 2010).

There have been various other health and environmental concerns expressed about CFLs, including allegations that CFLs may trigger migraine headaches, exacerbate skin problems, disrupt the body's production of melatonin, increasing "electric smog," increasing epileptic seizures, and releasing potential cancer-causing chemicals such as phenol, naphthalene, and styrene (Adams, 2011; Ward, 2011). Although these allegations have not been validated or confirmed, they have increased public anxiety and concerns about CFLs, feeding on the public's normal fear of the different and unknown.

Very little, if any, consideration was given to the significance of these health and environmental issues and how they might be best managed when the EISA was passed in 2007. This environmental concern has been described as a "sleeper" issue with the potential to create even greater media and public backlash against CFLs once this potential problem becomes better known (APT, 2010).

Another problem currently submerged but likely to flare up in the next couple years has to do with the timing of the legislative deadlines and the availability of substitute products. In the period 2012–2014, the initial compliance period after the phaseout of most incandescent bulbs commences, the two most available replacement choices will be halogen bulbs and CFLs. Earlier halogen bulbs produced substantial consumer concern due to the large amount of heat they generated and widely broadcast stories of fires started by halogen bulbs (APT, 2010). Newer halogen bulbs have much lower heat output and fire hazard and have the advantage of looking similar to the traditional incandescent bulb (APT, 2010). Yet, the halogen bulb is only about 25 percent more efficient than the incandescent bulb compared to the CFL's 75 percent efficiency improvement (APT, 2010), and thus the CFL is the preferred replacement technology from an energy efficiency perspective.

Another technology looming on the horizon is light-emitting diodes (LEDs), which many experts expect will be the longer-term solution to the lighting problem (Azevedo *et al.*, 2009). They have been used in decorative and other specialized applications for many years and are just beginning to become available for residential use, although they are currently very expensive. As the price of LEDs comes down and the quality goes up over the next few years, LEDs are expected to become the predominant lighting technology by the middle to the end of the decade (Azevedo *et al.*, 2009).

The legislative dictate to switch lighting sources in the 2012–2014 period is therefore problematic, given that the superior technology (LEDs) will not be widely available in this transition period, but are waiting in the wings to assume a dominant role a few years later. Consumers will be forced to switch from their long-standing reliance on the incandescent bulb to an unfamiliar technology (CFLs) for only a short period before switching again a few years later to LEDs. Back-to-back disruptive changes are sure to create significant public confusion and frustration. It also creates a dilemma for manufacturers—it is not very appealing to invest in the R&D, the costs of building manufacturing capacity, and the building of supply and distribution chains for a product that will only be in the ascendency for a couple years. Accordingly, there are predictions that there will be significant shortages in supply of general service lightbulbs in the United States in the 2014–2015 period when the *de facto* ban on incandescent lightbulbs is fully implemented (APT, 2010).

The combined effect of these factors has provoked a political backlash against the legislative phaseout of the incandescent bulb, with the perceived shortcomings of the CFL serving as a key catalyst to much of the controversy and opposition. Some consumers are stockpiling incandescent bulbs (Rice, 2011), in many cases after trying and rejecting CFLs. For example, Professor Glenn Reynolds, author of the highly popular InstaPundit blog, wrote: "I'm deeply, deeply disappointed with CFL bulbs. I replaced pretty much every regular bulb in the house with CFLs, but they've been failing at about the same rate as ordinary long-life bulbs, despite the promises of multi-year service. And I can't tell any difference in my electric bill. Plus, the Insta-Wife hates the light" (Reynolds, 2011).

Public resistance to the switchover is likely to grow as more consumers become aware of the legislative consequences as they begin to take effect on January 1, 2012. Some politicians have decried the "lightbulb ban," with one Congressman lambasting the legislation for involuntarily imposing on the public the harsh glare of those "little, squiggly, pigtailed" CFLs (Rice, 2011). The CFL has become a focus for broader political debates and divisions in the country about the role of government in society and consumer freedom.

Strongly worded rhetoric against the legislation is common, such as the following commentary published by *Bloomberg*: "The bulb ban makes sense only one of two ways: either as an expression of cultural sanctimony, with a little technophilia thrown in for added glamour, or as a roundabout way to transfer wealth from the general public to the few businesses with the know-how to produce the lightbulbs consumers don't really want to buy" (Postrel, 2011). Several bills have been introduced in Congress seeking to overturn the EISA requirements, which have garnered significant (in some cases majority) support, but not have succeeded in repealing the legislation to date. The House passed an amendment in July 2011 that would prohibit the Department of Energy from spending any funds on implementing the lighting efficiency standards, but the Senate had yet to act on this amendment at the time of this writing (Howell, 2011). In addition, similar bills have been introduced in state legislatures, and in July 2011 Texas became the first state to implement legislation relating to the lighting restrictions, passing a law stating that incandescent bulbs manufactured in Texas do not involve interstate commerce and thus are not subject to the federal restrictions (Simon, 2011).

One again, the government phaseout of incandescent bulbs in favor of CFLs may be justified on substantive grounds, but in terms of how it has been implemented has involved a growing controversy and has the potential to bloom into a full-fledged fiasco in the next couple years. A little more foresight and planning could have helped address or at least mitigate many of the concerns and controversies about the lightbulb switchover.

## Conclusion

Attempts by governments to induce beneficial technology change are a difficult and hazardous undertaking (Marchant, 2009; Nelson & Langlois, 1983). The three technology choices discussed in this paper represent socially beneficial technologies that governments legitimately decided were needed and market forces seemed unlikely to deliver anytime soon. Yet, the attempt to mandate these technologies proved to be enormously difficult, complex, and controversial in all three cases. In one case (ZEV mandate) the mandate failed, in another (DTV) the mandate ultimately succeeded, and in the third

(CFLs) the jury is still out. But notwithstanding the final outcome, all three attempts were plagued by opposition, delays, unanticipated impacts, and controversies. One lesson from this analysis is that technology mandates are difficult—predicting the timing and course of future technologies is complex and uncertain, and so trying to fix technology availability by regulatory fiat is prone to problems. Such governmental failures in trying to pick technologies and timelines in advance, enforced through rigid requirements that provided little or no flexibility if the initial assumptions turned out to be flawed, demonstrates the risks and challenges of such government mandates and suggests they should only be employed as a last resort when no other regulatory or voluntary approach is feasible.

The more central lesson of this analysis is that when governments decide to impose a technology mandate, they need to be more deliberate and thoughtful about the economic, environmental, social, stakeholder, and consumer implications and responses to the mandate.

For example, the EV mandate adopted by California and other states did not consider the technological limitations and feasibility of essential complementary technologies (i.e., advanced batteries), the performance features of such vehicles across the range of environments the vehicles would operate in, the public reticence to buy a new, unproven type of vehicle, and the need for building a recharging infrastructure. The *de facto* ban on incandescent lightbulbs did not consider potential mercury exposure issues associated with replacement CFL lightbulbs, the performance aspects of CFL lightbulbs that would annoy some consumers, and the development cycles of CFL vs. LED as a replacement lighting source. The mandate for DTV did not adequately consider the distributional impacts of this transition on low-income and rural citizens, the incentives for equipment manufacturers to continue producing analog TV sets, and the demand for converter equipment. In these and other cases, the government's failure to fully anticipate and plan to mitigate the variety of environmental, health, social, and economic impacts that the technology mandate would induce substantially undermined the efficacy of and support for the intended technology transformation. A more multifactorial assessment of the technology across the life span of the technology, considering technological, environmental, competitive, social, and economic implications would have helped to prevent some of the controversies and problems associated with the technology mandates and would almost surely have improved the implementation and success of the technology mandates.

Given this record of systemic failure to anticipate the full life cycle socio-behavioral implications and consequences of technology mandates, it would behoove a legislative or regulatory body considering the imposition of such a mandate to undertake a careful, detailed, foresight study in advance of the mandate. Such a study could be undertaken by a third party, such as the former Office of Technology Assessment, or it could be conducted internally. The analysis would serve two purposes. First, it could anticipate and try to prevent potential problems by factoring such issues in the design of the mandate. Thus, for example, CARB might have provided up front additional flexibility for meeting the ZEV mandate with other types of clean fuel vehicles such as hybrids, the DTV transition might have included a requirement for equipment manufacturers to phase out analog TV sets up front, and the incandescent lightbulb phaseout might have stretched out the timeline to take better advantage of the emerging LED technology.

The second reason for the up-front study would be to anticipate and prepare to address other problems. For example, the ZEV mandate might have better planned for the potential problem that battery performance and costs would not meet expectations, the DTV transition could have included a better-designed public education and coupon program up front, and the incandescent bulb phaseout could have included better preparation and infrastructure to ensure the recycling of mercury-containing CFL bulbs.

## References

Adams, S. (2011, January 31). Energy saving light bulbs 'could trigger breast cancer,' *The Telegraph (UK)*, available at http://www.telegraph.co.uk/health/healthnews/8288982/Energy-saving-light-bulbs-could-trigger-breast-cancer.html.

Angelle, A. (2010). Will LED light bulbs best your CFLs and incandescents? *Popular Mechanics*, Aug. 4, 2010, available at http://www.popularmechanics.com/science/environment/will-led-light-bulbs-best-cfls-and-incandescents"led-light-bulbs-best-cfls-and-incandescents.

Antonoff, M. (2007). Digital TV at last?, *Scientific American*, February, 70–85.

Applied Proactive Technologies, Inc. (APT) (2010), The U.S. Replacement Lamp Market, 2010-2015, and the Impact of Federal Regulation on Energy Efficiency Lighting Programs (Aug. 2010).

Aucott, M., Michael, M., & Michael, W. (2003). Release of Mercury from Fluorescent Bulbs, *J. Air & Waste Management Association, 53,* 143–151.

Azevedo, I. L., Morgan, M. G., & Morgan, F. (2009). The Transition to Solid-State Lighting, *Proceedings of the IEEE, 97,* 481–510.

Battery Technical Advisory Panel (1995). Performance and Availability of Batteries for Electric Vehicles, Report prepared for the California Ai Resources Board, Dec. 11, 1995.

Bedsworth, L. W., & Taylor, M. R. (2007). Learning from California's Zero-Emission Vehicle Program, *California Economic Policy* (Public Policy Institute of California), Vol. 3(4), 1–20 (Sept. 2007).

Bohan, S. (2011, April 7). Mercury in unrecycled CFLs takes a toll on the environment. *Los Angeles Times,* p. 4.

Boyce, J. M. (2009). The U.S. Digital Television Broadcasting Transition. *IEEE Signal Processing Magazine,* May, 109–112.

Burke, A. F., Kurani, K. S., & Kenney, E. J. (2000). Study of the Secondary Benefits of the ZEV Mandate (Prepared for CARB, Aug. 2000).

Calfee, J. E. (1985). Estimating the Demand for Electric Automobiles Using Fully Disaggregated Probabilistic Choice Analysis, *Transportation Research B, 19B,* 287–301.

California Air Resources Board (CARB) (1990). Proposed Regulations for Low-Emission Vehicles and Clean Fuels: Staff Report (Aug. 13, 1990).

California Air Resources Board (CARB) (1991). Proposed Regulations for Low-Emission Vehicles and Clean Fuels: Final Statement of Reasons (July 1991).

California Air Resources Board (CARB) (1994a). Draft Technical Document for the Low-Emission Vehicle and Zero-Emission Vehicle Workshop on March 25, 1994: Zero-Emission Vehicle Update (Mar. 2, 1994).

California Air Resources Board (CARB) (1994b). Staff Report: 1994 Low Emission Vehicle and Zero-Emission Vehicle Program Review (Apr. 1994).

California Air Resources Board (CARB) (1996a). Board Resolution 96-12, at 5–6 (Mar. 28, 1996).

California Air Resources Board (CARB) (1996b). Final Statement of Reasons for Rulemaking Including Summary of Comments and Agency Response (Mar. 28, 1996).

California Air Resources Board (CARB) (1996c). Notice of Public Hearing to Consider Amendments to the Zero-Emission Vehicle Requirements for Passenger Cars and Light-Duty Trucks (Mar. 28, 1996).

California Air Resources Board (CARB) (2000). Staff Report: 2000 Zero Emission Vehicle Program Biennial Review (Aug. 7, 2000).

California Air Resources Board (CARB) (2003, April 24). ARB modifies zero emission vehicle regulation [News Release].

California Air Resources Board (CARB) (2004). Final Statement of Reasons: The 2003 Amendments to the California Zero Emission Vehicle Program Regulations (Jan., 2004).

California Air Resources Board (CARB) (2009). White Paper: Summary of Staff's Preliminary Assessment of the Need for Revisions to the Zero Emission Vehicle Regulation (Nov. 25, 2009).

Consumer Electronics Ass'n v. Federal Elections Commission (CEA v. FCC), 347 F.3d 291 (D.C. Cir. 2003).

Consumer Product Safety Commission (CPSC) (2010). Trisonic Compact Fluorescent Light Bulbs Recalled Due to Fire Hazard, CPSC Release #11-001, Oct. 5, 2010, available at http://www.cpsc.gov/cpscpub/prerel/prhtml11/11001.html.

Department of Energy (DOE) (2011, July 18). Energy savers: When to turn off your lights, available at http://www.energysavers.gov/your_home/lighting_daylighting/index.cfm/mytopic=12280.

Di Maria, C., Ferreira, S., & Lazarova, E. (2010). Shedding Light on the Light Bulb Puzzle: the Role of Attitudes and Perceptions in the Adoption of Energy Efficient Light Bulbs, *Scottish Journal of Political Economy, 57,* 48–67.

Dunbar, J. (2008, January 2). Feds share coupons to help TV transition. *Associated Press.* Retrieved from http://www.ap.org/

Energy Star (2010). *Frequently asked questions information on compact fluorescent light bulbs (CFLs) and mercury*, November 2010, available at http://www.energystar.gov/ia/partners/promotions/change_light/downloads/Fact_Sheet_ Mercury.pdf.

Environmental Protection Agency (EPA) (2009, June 11). The digital television transition: Don't kick your TV to the curb – eCycle instead [News Release].

Environmental Protection Agency (EPA) (2011, January 25). *What to do if a compact fluorescent light (CFL) bulb or fluorescent tube light bulb breaks in your home*, available at http://www.epa.gov/cfl/cflcleanup.pdf.

Farhi, P. (2009, June 10). Broadcast TV Never converted its digital dream. *Washington Post.*

Government Accountability Office (GAO) (2005). *Digital television transition: Issues related to an information campaign regarding the transition.* Washington, DC: GAO

Government Accountability Office (GAO) (2007). *Digital television transition: Increased Federal planning and risk management could further facilitate the DTV transition.* Washington, DC: GAO

Gruenspecht, H. (2001). Zero emission vehicles: A dirty little secret. *RFF Resources,* 7–10.

Hart, J. A. (2010). The transition to digital TV in the United States: The endgame. *International Journal of Digital Television, 1,* 7–29.

Hart, K. (2009a, January 9). Obama urges delay in digital TV switch. *Washington Post,* p. D1.

Hart, K. (2009b, June 11). Digital TV ready to rule the tube; leaving some viewers in the dark, *Washington Post,* p. A1.

Hart, K. (2009c, January 14). Turmoil over TV switch grows. *Washington Post,* p. D1.

Hart, K. (2008a, September 17). Lawmakers see challenges for TV transition. *Washington Post,* p. D2.

Hart, K. (2008b, May 20). Digital TV transition not as easy as advertised. *Washington Post,* p. A1.

Hart, K. & Whoriskey, P. (2009, February 14). Stalled switch to digital TV a classic tale of breakdown. *Washington Post,* p. A1.

Hazlett, T. W. (2001). Testimony before the Senate Committee on Commerce, Science, and Transportation, Hearings on the Transition to Digital Television Broadcasting, March 1, 2001.

Howell, K. (2011, July 15). House votes to block funding for light bulb standards. *Greenwir,* available at http://www.eenews.net/gw/2011/07/15/.

Hurst, T. (2011, April 11). Is the electric car's tax-free ride coming to an end? *Reuters,* available at http://www.reuters.com/article/2011/04/11/idUS175572720320110411.

Jaffe, A. B., Richard, G. N., & Robert, N. S. (2002). Environmental policy and technology change. *Environmental & Resource Economics, 22,* 41–69.

Jaffe, A. B., & Stavins, R. N. (1994). The energy paradox and the diffusion of conservation technology. *Resource and Energy Economics, 16,* 91–122.

J.D. Power and Associates (2010). *Drive Green 2020: More hope than reality,* available at http://businesscenter.jdpower.com/JDPAContent/CorpComm/pdfs/DriveGreen2020_102610.pdf.

King, C. W., & Webber, M. E. (2008). The Water Intensity of the Plugged-In Automotive Economy, *Environmental Science & Technology* 42, 4305–4311.

KRC Research (2010, December 15). 3rd Annual Sylvania Socket Survey, available at http://assets.sylvania.com/assets/Documents/2009_SYLVANIA_Socket_Survey.d81a552e-cb6b-4779-9e56-5da47e838c7f.pdf.

Labaton, S. (2002, April 29). Most Commercial broadcasters will miss deadline for digital television. *New York Times*, p. A16.

Labaton, S. (2009, June 6). Millions face blank screens in TV switch, *New York Times*, pp. A1, B2.

Lighting Research Center (LRC) (2003). Increasing Market Acceptance of Compact Fluorescent Lamps (CFLs). Report prepared for the U.S. Environmental Protection Agency, Sept. 30, 2003.

Logan, J. (2008). Lighting Efficiency Standards in the Energy Independence and Security Act of 2007: Are Incandescent Light Bulbs "Banned"?, CRS Report RS22822, April 23, 2008.

Magat, W. A. (1979). The effects of environmental regulation on innovation. *Law and Contemporary Problems, 43,* 3–25.

Marchant, G. E. (2009). Sustainable energy technologies: Ten lessons from the history of technology regulation. *Widener Law Journal, 18,* 831–858.

Massachusetts Institute of Technology (MIT) (2010). Electrification of the Transportation System: An MIT Energy Initiative Symposium, April 8, 2010.

National Research Council (NRC) (2006). *State and Federal standards for mobile-source emissions.* Washington, DC: National Academies Press.

National Research Council (NRC) (2010). *Hidden costs of energy: Unpriced consequences of energy production and use.* Washington, DC: National Academies Press.

Nelson, R. R., & Richard, N. L. (1983). Industrial innovation policy: Lessons from American history. *Science, 219,* 814–818.

That Newfangled Light Bulb [Editorial]. (2008, February 17). *The New York Times,* available at http://www.nytimes.com/2008/02/17/opinion/17sun3.html.

Palm, E. (2009, June 9). DTV transition: Avoiding an e-waste "Tsunami." *CNET News*, available at http://news.cnet.com/8301-11128_3-10260174-54.html.

Pegoraro, R. (2009, January 22). TV's digital transition doesn't have to cause headaches. *Washington Post*, p. D3.

Polakovic, G. (2002, September 15). State takes sharp turn on emissions; cars: With electric vehicles still impractical, hybrids and gasoline engines are showing unexpected progress (quoting CARB Chairman Alan C. Lloyd). *Los Angeles Times*, Part 2, p. 1.

Postrel, V. (2011, June 10). Need a Light Bulb? Uncle Sam Gets to Choose. *Bloomberg View*, available at http://www.bloomberg.com/news/2011-06-10/need-a-light-bulb-uncle-sam-gets-to-choose-virginia-postrel.html.

Ramsay, M. (2010). Environmental lift of battery cars is limited. *Wall Street Journal*, Oct. 18, 2010.

Reynolds, G. (2011). Instapundit.com, May 27, 2011, available at http://pajamasmedia.com/instapundit/121431/.

Ricardo (2011). Preparing for a Life Cycle $CO_2$ Measure, May 20, 2011, available at http://urbact.eu/fileadmin/Projects/EVUE/documents_media/Preparing_for_a_Life_Cycle_CO2_Measure.pdf.

Rice, A. (2011). Bulb in, bulb out. *New York Times Magazine*, June 3, 2011.

Sandahl, L.J., *et al.* (2006). Compact Fluorescent Lighting in America: Lessons Learned on the Way to Market, Report to DOE Prepared by Pacific Northwest National Laboratory (May 2006), available at http://apps1.eere.energy.gov/buildings/publications/pdfs/ssl/cfl_lessons_learned_web.pdf.

Scelfo, J. (2008, January 10). Any other bright ideas? *New York Times*, available at http://www.nytimes.com/2008/01/10/garden/10lighting.html?pagewanted=all.

Scientific Committee on Health and Environmental Risks (SCHER) (2010). Opinion on Mercury in Certain Energy-saving Light Bulbs, European Union Health & Consumer Protection Directorate-General, May 18, 2010.

Senate Rep. No. 403, 90th Cong, 1st Sess. (1967).

Shiver, J. (2001, July 30). Broadcasters Seek to Delay Digital TV. *Los Angeles Times*, available at http://articles.latimes.com/2001/jul/30/business/fi-28189.

Silveira, G. T. R., & Chang, Shoou-Yuh (2011). Fluorescent lamp recycling initiatives in the United States and a recycling proposal based on extended producer responsibility and product stewardship concepts. *Waste Management & Research, 29,* 656–668.

Simon, R. (2011, July 9). Texas aglow with effort to save the incandescent bulb. *Los Angeles Times,* available at http://articles.latimes.com/2011/jul/09/nation/la-na-adv-texas-light-bulbs-20110710.

Smith, R. (2011). The new light bulbs lose a little shine. *Wall Street Journal,* Jan. 19, 2011.

Sperling, D., & Gordon, D. (2008). Advanced passenger transport technologies. *Annual Review of Environment and Resources, 33,* 63–84.

Steinberg, J. (2007, June 7). Converters signal a new era for TVs. *New York Times,* p. C3.

Stelter, B. (2009a, June 14). Changeover to digital TV off to a smooth start. *New York Times,* p. A16.

Stelter, B. (2009b, June 18). Switch to digital TV broadcasts is a work in progress. *New York Times,* p. B8.

Stewart, R. B. (1981). Regulation, innovation, and administrative law: A conceptual framework. *California Law Review, 69,* 1256–1270.

Swope, T. (2010). The Present and Possible Future CFL Market, Presentation to the Northeast Residential Lighting Stakeholders Meeting on behalf of D&R International for US DOE, March 30, 2010, available at http://neep.org/uploads/Summit/2010%20Presentations/NEEP%20Lighting_Swope.pdf.

Taub, E. A. (2002, September 12). The big picture on digital TV: It's still fuzzy. *New York Times,* pp. E1, E7.

Taub, E. A. (2008, December 22). In move to digital TV, confusion is in the air. *New York Times,* p. B5.

The Year 2000 Battery Technology Advisory Panel, Advanced Batteries for Electric Vehicles: An Assessment of Performance, Cost, and Availability (Prepared for CARB, 2000).

Waide, P. (2010). Phase Out of Incandescent Lamps: Implications for International Supply and Demand for Regulatory Compliant Lamps, International Energy Agency Information Paper (April 2010).

Ward, V. (2011, April 20). Energy saving light bulbs 'contain cancer causing chemicals.' *The Telegraph (U.K.),* available at http://www.telegraph.co.uk/health/8462626/Energy-saving-light-bulbs-contain-cancer-causing-chemicals.html.

# Chapter 5

# Issues in Life Cycle Risk Assessment: The Way Ahead

**Gurumurthy Ramachandran**
*Division of Environmental Health Sciences, 420 Delaware Street SE, MMC 807, Minneapolis, MN 55455*
ramac002@umn.edu

## Introduction

Nanotechnology is the manufacturing and application of materials and devices at the nanoscale (1–100 nm) by using unique characteristics of nanoparticles (NP), which are different than those of larger particles. Nanoparticles are defined as particles greater than 1 nm and smaller than 100 nm in two or three dimensions (ISO, 2007). At this size scale, particles exhibit physicochemically unique optical, magnetic, and electrical characteristics that find many uses in technology (Kreyling *et al.*, 2006).

The rapid growth of nanotechnology in recent years encompasses a range of industries, including pharmaceuticals, materials, medicine, agriculture, electronics, and energy (Lane & Kalil, 2005). New products have emerged from the laboratories and into the worldwide commercial market estimated to be as large as $1 trillion

*Emerging Technologies: Socio-Behavioral Life Cycle Approaches*
Edited by Nora Savage, Michael Gorman, and Anita Street

ISBN 978-981-4411-00-4 (Hardcover), 978-981-4411-01-1 (eBook)
www.panstanford.com

by 2015 (Roco, 2007). On the basis of the promise of the technology to lead to new jobs and economic growth, there has been significant investment in this technology by governments and industry worldwide (Lux Research, 2007; Roco, 2007).

As consumer products that contain nanomaterials become more commonplace, it is only a matter of time before a significant proportion of the population will use or come into contact with products containing nanomaterials. However, the earliest and potentially the most significant exposures and risks will likely be in the occupational arena. An estimated two million new workers will be exposed to engineered nanomaterials in occupational environments over the next 15 years (Roco, 2003). There are several industry sectors and processes where worker exposures to nanomaterials have the potential to be significant if not properly contained, including chemical and pharmaceutical companies, construction and manufacturing (e.g., powder handling & cement), and electronics and communications. Despite the large investments in nanotechnology, corresponding investments in environmental, health, and safety aspects of this technology and its processes and products have not been as high.

Much is still unknown or poorly known regarding the health risks of nanomaterials. For example, the mass concentration has traditionally been the metric for exposure assessment of airborne particles and is the basis for regulation even for nanomaterials such as the recently proposed NIOSH Recommended Exposure Limit (REL) for carbon nanotubes (CNTs) of 7 $\mu g/m^3$. However, alternative metrics such as surface area and number concentration have been proposed for characterizing nanoparticle exposures. Thus, uncertainty pervades even basic issues.

Given the predicted far-reaching influence of nanotechnology over our lives, the dynamic nature of many engineered nanomaterials, and the lack of knowledge of human health and ecological risks relating to this technology, a life cycle approach to considering the risks relating to nanomaterials is prudent. The potential impacts at every stage of a material's life cycle—from production to transport to use to end-of-life treatment and disposal/recycling need to be studied (Klöpffer *et al.*, 2007). Such analyses may affect the design of products based on these technologies. Life cycle analysis includes a consideration of the material and energy flows involved in the manufacture and commerce of nanotechnology-based products, important environmental impacts, burdens, and weak spots, thereby identifying the relative contributions from each life cycle stage. Most studies to date have focused on inventory analysis and material and energy flows (Lave *et al.*, 1995). Lloyd and Lave (2003) studied the economic and environmental implications of using nanocomposites in automobiles, while Lloyd *et al.* (2005) studied the relative benefits of using nanotechnology to stabilize platinum metal particles in automotive catalyst technologies. However, none of these studies considered occupational and environmental risks in their calculations.

Risk assessment is an analytical approach, which helps determine whether the level of a hazard and its probability of occurrence are acceptable or if the risk needs to be managed. Life cycle risk analysis integrates these two ideas to address issues such as the likelihood and magnitude of risk contribution from each stage of the product's life cycle, the availability of data for risk assessments, and the options for managing the risk. Where potential impact depends on physical form as well as chemistry, changes in physicochemistry—along with availability or exposure potential—across a material's life cycle can have a profound impact on risk within different contexts.

Within this complex challenge, much attention has been placed on exposure potential as a first order determinant of potential risk. Kohler *et al.* (2008) studied the potential release of carbon nanotubes throughout the life cycle of two different products as case studies—lithium ion secondary batteries and synthetic textiles. They found that release of CNTs can occur in not only the production phase but also the usage and disposal phases. For example, textile production, where CNTs are used as additives, has several scenarios that can lead to potential exposures—blending of polymers with CNTs that can involve dry powders or liquid phase dispersion and fabric finishing and tailoring that may involve wet or dry abrasion of fiber fragments. In the use phase, degradation of the matrix in composite textiles can result in CNT release, because CNTs are much more stable and do not degrade in the same manner and time as polymers. Textile industrial waste from fabric manufacture is recovered and reused often. In fiber-to-fiber recycling, either the material is physically shredded or the textiles are cut into small pieces and granulated to form pellets. All these mechanical and thermal processes provide opportunities for CNT release. Disposal could be through landfilling or incineration. While municipal solid waste incinerators most likely will completely incinerate CNT/polymer composites, uncontrolled incineration in open fires that occur frequently in developing countries will lead to emissions of CNTs as it is an incomplete combustion process. Thus, the risk presented by just-generated carbon nanotubes, for instance, may be markedly different from the risk presented by processed/purified nanotubes, which represent not only an altered physicochemistry but also a different exposure potential. Likewise, once these carbon nanotubes have been incorporated into a product—for example, a fabric composite—the exposure potential and the physicochemical nature of any material that is released are profoundly different from that of the starting material. As the resulting product is used and eventually disposed or recycled, the hazard and exposure potential differ yet again.

Thus, the risk profile of a nanomaterial over its life cycle is complex, even if that material is relatively stable. However, when nanomaterials undergo transformations through their life cycle through processes, such as agglomeration, dissolution, surface adsorption/desorption, chemical reaction, or other interactions with close-proximity materials, the challenges of evaluating and addressing risk become more difficult. We have little quantitative information about the exposure potential in each of the phases of the life cycle described above. The populations being exposed in each of these phases also differ as, presumably, do their susceptibilities.

In summary, a life cycle risk assessment requires the estimation of risk at every step of the life cycle of the product and, therefore, an assessment of the hazard and exposure at every step of the life cycle. The following sections describe the risk assessment process, the roadblocks to carrying this out for nanomaterials, and the potential ways to move ahead.

## Traditional Risk Assessment

Risk assessment is a complex process that involves the integration of hazard, exposure, and health effect information to characterize the potential for risk to human health (Kandlikar *et al.*, 2007). This requires information across a range of domains, including source characterization, fate and transport, modeling, exposure assessment, and dose-response characteristics. Such methods typically utilize quantitative predictions of health impacts and explicitly model and incorporate uncertainties.

Modern risk assessment therefore aims to present decision makers with a range of options, so that they may decide on the protective policies that are warranted in light of the range of possible future outcomes of alternative policies (Haimes & Lambert, 1999). If there is uncertainty regarding exposures or dose in a population, for example, risk assessors can decide to collect more data, extrapolate values from other similar populations, or use numerical models to estimate missing values.

The various elements of the risk assessment paradigm incorporate factors relating to exposure and toxicity. Exposure is defined as the intensity of contact between a contaminant and the relevant biological sites of impact over a relevant period of time. Exposure is determined by assessing sources of pollutants and their strengths, measuring or modeling concentrations in environmental media, measuring or modeling human exposures through various pathways, and sometimes through biological monitoring to measure tissue burden to determine dose. The estimation of a biologically relevant dose from exposure information is, however, often very difficult and requires fairly detailed knowledge of the toxicokinetics of the pollutant in the human body. The effects of exposure and biologically relevant dose can be estimated on the basis of the information provided by *in vitro* and *in vivo* studies, QSAR modeling, and epidemiological studies.

By quantifying exposures and effects, risk assessors can determine the extent of risk and choose appropriate measures for managing risks. Implicit in this process is the setting of standards or guidelines regarding safe or acceptable levels of exposure for a population. For non-cancer toxicants, it is often assumed that there is some level below which there are no adverse effects—the no observed adverse effect level (NOAEL). An acceptable exposure limit is established below this threshold. For carcinogens, the standard practice is to assume that no threshold exists below which there is no risk to human health (i.e., the threshold is zero). Exposures exceeding the prescribed threshold are considered to cause adverse effects, and measures should be taken to mitigate or reduce exposures. Health risks can be calculated for different exposures if a "dose-response" curve is well defined above this exposure threshold. Dose-response curves are typically extrapolated from high to low dose and are assumed to be linear. The excess risk is calculated by multiplying the dose by the dose-response-curve slope factor. Though this health-risk model assumes no threshold level, for the purpose of risk management and prioritization, risks exceeding some minimum risk probability (e.g., $1/10^3$ for occupational populations or $1/10^6$ for non-occupational populations) are considered to be of concern.

However, the standard approach described above cannot, at present, be easily applied to most nanomaterials. This is because key mechanisms for exposure processes and toxicity effects of engineered nanomaterials remain poorly understood. Uncertainties include those related to such general questions as: (a) How long do manufactured nanomaterials persist in the atmosphere? (b) How stable are nanomaterials over time given specific occupational conditions? (c) What is the effect of particle shape on their fate and transport? (d) What are likely routes of exposure (e.g., inhalation, dermal, ingestion, & ocular)? (e) What are the metrics by which exposure should be measured (e.g., particle mass or number or surface area concentration)? (f) What are key mechanisms of translocation to different parts of the body after nanomaterials enter the body? (g) What are the possible mechanisms of toxicity, including oxidative stress due to surface reactivity, presence of transition metals leading to intracellular calcium and gene activation, and intracellular transport of nanomaterials to the mitochondria? (Kandlikar *et al.*, 2007). In assessing overall risk, both the exposure and hazard aspects of risk are poorly understood.

The following sections will describe the challenges in (a) toxicity assessment, (b) exposure assessment, and (c) risk modeling. Some of these challenges arise from the oversight mechanisms and legal structures in place, which affect the ease with which some of the needed information can be obtained, while others arise from difficulties in developing accurate toxicity assays, while still others arise from the analytical frameworks for assessing risks, which may not always be capable of addressing novel risks.

## Challenges and Opportunities Presented by Novel Technologies

### Legal Limitations

In the United States, regulation of the human health risks is primarily the responsibility of the Occupational Safety and

Health Administration (OSHA) for occupational risks and the US Environmental Protection Agency (EPA) for non-occupational risks under the OSH Act and the Toxic Substance Control Act (TSCA), respectively. A series of court rulings has resulted in an overly burdensome process of health risk assessment for any substance being placed on the under-resourced OSHA (Ashford & Caldart, 1996). For example, the so-called benzene and cotton dust decisions by the US Supreme Court in 1980 and 1981, respectively, mandated that before proposing any new standard, OSHA should conduct a complete risk assessment, prove that a significant risk for health exists, prove that the proposed standard will increase worker protection, and also prove that the new standard is technologically and economically feasible. This has led to a standard setting process that is so slow that thousands of chemicals have no defined occupational exposure limits (OELs). Any new nanomaterial would likely meet the same fate (Balbus *et al.*, 2007).

A similar paralysis besets the oversight of chemicals in the environment by TSCA, where the EPA faces a logical catch-22 (Chatterjee, 2008a; 2008b). Before the EPA can ask the producer to provide data to help in risk assessment of a chemical, the agency needs to show that the chemical presents an unreasonable risk to human health or the environment. EPA thus needs toxicity and exposure data that producers are not obligated to provide unless the EPA can first show that a risk exists. Thus, only a miniscule fraction of the 62,000 chemicals existing in 1979 has been reviewed by the EPA (US GAO, 1994). High production volume (HPV) chemicals accounted for more than 90% of these; since 1979, the EPA has run a voluntary program whereby firms can provide screening-level data for these chemicals. Although such information has been submitted for most of these chemicals, it is not sufficient to support risk-based decision-making. For pre-manufacture notification (PMN) of new chemicals, TSCA only requires that producers submit toxicity testing information that is "in their possession" when they file the PMN; it does not require new testing. Not surprisingly, 85% of PMNs lack data on chemical health effects (US GAO, 2005).

Limitations of regulatory oversight are also evident in private sector R&D related to nanoparticles. Despite the large and rising investments in nanotechnology research, corresponding private investments in studying the health and safety aspects have been limited. Private industry is less willing to conduct long-term studies on chronic health and environmental effects (Lyndon, 1989). Wagner (2004) ascribed several reasons as to why the private sector is reluctant to voluntarily undertake long-term safety testing of their products. First, direct costs associated with assessing risks are expensive and may not produce definitive results, especially if tests have low specificity. Second, virtually no market benefits accrue to firms who produce research on long-term safety of products. As a result, firms do not consider the assessment of long-term risks from chemicals to be an attractive business investment.

The common theme running through the OSHA and TSCA examples is that the entire burden of data collection and risk assessment is placed on agencies without the budgetary means to carry out this mandate; while firms have little incentive to reveal toxicity or exposure information. EPA has taken the initial steps for nanoparticle regulation through the implementation of a voluntary nanotechnology stewardship program. In the light of oversight of existing chemicals in the US, it is reasonable to assume that the voluntary program will fail even as thousands of new nanoproducts enter the market in the coming decade (Bowman & Hodge, 2007). It is, therefore, important to consider alternative approaches that may show better performance in terms of the efficiency (using least resources) and effectiveness (extent of coverage of new nanomaterials) of producing information essential for risk assessment. In this vein, Ramachandrann *et al.* (2011a) proposed a framework of dynamic oversight that could (a) move dynamically between "soft" and "hard" approaches as information and nanoproducts evolve; (b) integrate inputs from all stakeholders, with strong public engagement in decision-making to assure adequate analysis and transparency; and (c) include an overarching coordinating entity to assure strong inter-agency coordination and communication that can meet the challenge posed by the convergent nature of nanobiotechnology.

## Toxicity

### Conservatism in toxicity testing: The costs

The wide spectrum of nanomaterials in manufacturing implies that exposure to nanomaterials can occur through inhalation, ingestion, and dermal and parenteral routes. Studies also indicate that the safety and toxicity of nanomaterials are different than those for the same materials on larger size scales with respect to composition and surface properties. The rapidity with which this technology is growing could push this sector from ~800 nanomaterials currently known to numbers in the 10,000 range within a decade. Thus, a re-consideration of the current toxicological paradigm is needed so as to avoid the current situation of the chemical industry, where among the ~40,000 industrial chemicals, fewer than 1000 have undergone toxicity testing. One of the major factors contributing to this backlog is the high cost and length of time to complete even a single toxicological screen through animal testing. Conservative methods for the toxicological assessment for each new nanomaterial are expensive in terms of time, laboratory animals, materials, and expenditures.

There is no consensus about how to proceed with nanomaterial toxicity testing, the toxicological end points to screen for, the comprehensiveness of the screening effort, the correct balance of *in vitro* (cellular & molecular) versus *in vivo* (animal or whole organism) testing, the cost of the effort, and who should be responsible for screening and safety assessment of nanomaterials.

Choi *et al.* (2009) analyzed the impact of testing costs on society's ability to gather information about nanoparticle toxicity and whether such costs can reasonably be borne by an emerging industry or the government.

Table 5.1 shows for the United States that costs for testing existing nanoparticles range from $249 million for optimistic assumptions about nanoparticle hazards (i.e., they are primarily safe and mainly require simple screening assays) to $1.18 billion for a more comprehensive precautionary approach (i.e., all nanomaterials require long-term *in vivo* testing). At mid-level estimates of total corporate R&D spending, assuming current levels of spending on hazard testing, the time taken to complete testing is likely to be very high (34–53 years) if all *existing* nanomaterials are to be thoroughly tested (conservative approach). These delays will only increase with time as new nanomaterials are introduced. The delays are considerably less if less stringent yet risk-averse perspectives are used. Choi *et al.* (2009) support a tiered risk-assessment strategy similar to the EU's REACH legislation for regulating toxic chemicals. Tiered strategies begin with an initial screening tier comprised of relatively simple and inexpensive tests; the outcomes of simple tests are used to prioritize substances for further, more resource-intensive and complex testing with increasing degrees of selectivity for adverse effect (Becker *et al.*, 2006). Such a system might be able to fill the enormous data gap for untested nanomaterials.

**Table 5.1** Estimated time (in years) needed to assess the hazards from existing nanoparticles at current R&D spending levels. Each entry is indexed by spending on nano-hazard research (10%, 5%, 1% of sectoral R&D) as well as a subjective perspective (optimistic, neutral, risk-averse, precautionary) on the distribution of risk levels from nanoparticles across particle types.

| | R&D Scenarios | | | | | | | | |
|---|---|---|---|---|---|---|---|---|---|
| | Scenario I | | | Scenario II | | | Scenario III | | |
| Spending on Hazards Research (%) | **10%** | **5%** | **1%** | **10%** | **5%** | **1%** | **10%** | **5%** | **1%** |
| **Optimistic** | 0.8 | 1.5 | 7.6 | 1.1 | 2.2 | 11.1 | 0.7 | 1.4 | 7.1 |
| **Neutral** | 1.5 | 3.1 | 15.3 | 2.2 | 4.5 | 22.3 | 1.4 | 2.9 | 14.3 |
| **Risk-Averse** | 2.3 | 4.6 | 23.2 | 3.4 | 6.7 | 33.7 | 2.2 | 4.3 | 21.6 |
| **Precautionary** | 3.6 | 7.3 | 36.5 | 5.3 | 10.6 | 53.0 | 3.4 | 6.8 | 34.0 |

*Source:* Reproduced with permission from Choi et al. (copyright 2009, American Chemical Society)

## Limitations of current *in vitro* methods in nanoparticle risk assessment

*In vitro* tests are used to evaluate the generic toxicity of substances and are primarily used as a screening tool as well as a guide to the proper selection of an appropriate animal model for *in vivo* testing. They are cost-effective and faster in assessing toxicity end points. However, studies to date have been somewhat contradictory in their assessment of the correlation between *in vitro* and *in vivo* tests (Donaldon *et al.*, 2008; Sayles *et al.*, 2007). While there is obviously a great need for relevant and accurate *in vitro* tests for nanotoxicity testing, current methods for such testing have additional serious limitations. These tests were established for assessing toxicity of conventional chemicals and are not appropriate for assessing the hazards of nanoparticles with novel properties. Nanoparticle properties may cause them to interfere with assay ingredients and detection systems and cause artifacts in cytotoxicity studies. These properties include (a) high-adsorption capacity that could cause the adsorption of nutrients and growth factors from culture media and is dependent on surface charge and hydrophobicity, (b) light absorptive nature of metallic properties that can affect cell viability readouts, (c) catalytic activity enhanced by the high surface area/mass ratios that can increase the production of reactive oxygen species and affect assays based on substrate oxidation, (d) magnetic properties that can affect production of free radicals (Kroll *et al.*, 2009). Therefore, *in vitro* testing requires a complete characterization of nanoparticle properties using reference materials and validation of assay techniques. The health effects or surrogates can also be extrapolated from studies performed on similar populations to determine toxicity parameters that may not be otherwise available. However, in the case of emerging technologies, it may be difficult to find comparable data from which parameters can be estimated, and uncertainty factors are typically required to account for the errors that are introduced during extrapolation (Kuempel *et al.*, 2007). This becomes especially significant when risk models use default distributional assumptions regarding important parameters. Such assumptions lead to severe risk underestimation, as described in a later section on "Fat Tail Distributions."

However, the past decade has witnessed the emergence of newer methods for toxicity testing in areas such as systems toxicological modeling, toxicogenomics, and high throughput toxicity screening. The following sections will briefly describe these developments.

### "Omics" and systems biology

Systems toxicology is an integrative approach to studying biological systems at the molecular level that combines data from mRNA, proteins, and other molecules with mathematical and statistical models. Toxicogenomics, toxicoproteomics, toxicometabolomics, transcriptomics, and bioinformatics play key roles in this new science. Toxicogenomics studies the manner in which the genome is involved in the response to toxic substances. Nanomaterials can interact with intracellular proteins, DNA, RNA, and other cellular molecules. The genomic approach has the potential to provide information about the mechanism of toxicity at the molecular level and identify global changes in the gene expression due to the effect of nanomaterial exposure (Jeong *et al.*, 2011). Oxidative stress, apoptosis, and altered immune function are some of the key mechanisms of toxicity. The protective antioxidant response activates the NF-E2-related factor

(Nrf)-2 signaling pathway to promote phase 2 promoter genes. However, beyond a certain level of oxidative stress, this response is overwhelmed by inflammation or apoptosis. Inflammation is initiated by pro-inflammatory signaling cascades, for example, the MAPK and NF-κB cascades, while apoptosis results from mitochondrial perturbation and release of pro-apoptotic factors (Nel *et al.*, 2006). Toxicoproteomics and metabolomics help in the understanding of nanomaterial–protein and protein–protein interactions. Gene and enzyme regulation ultimately affect proteins and metabolites and thus phenotypic expression. Thus, the elucidation of biological pathways and mechanism of toxicity becomes possible, enabling reduction of uncertainties in risk assessment.

The various "omics" techniques have the potential to allow the identification of the molecular target of a nanomaterial toxicant and define the dose–response relationship for the induction of toxicity. The reduction in uncertainties associated with various components of the risk assessment process (e.g., high-dose to low-dose extrapolation or animal-to-human extrapolation) has the potential to reduce the need for safety factors.

A significant potential benefit is that "omics" information will allow the prediction of the toxicity for specific nanomaterials within nanomaterial classes, along with higher study specificity, increased sensitivity in detecting risks, identification and quantification of susceptible populations, and identification of biomarkers of exposure and effect (Henry *et al.*, 2002). While these technologies have immense promise, there are also severe challenges to overcome before these promises are realized—(1) establishing a link between gene expression information and toxicologic effects, (2) distinguishing between adaptive responses and adverse effects, (3) availability of high quality, publicly available genomic and proteomic databases, and (4) determining background prevalence of gene-expression patterns in the population—to name just a few. Unless these challenges are overcome, there is a high likelihood for "omics" findings to be misinterpreted (Henry *et al.*, 2002).

George *et al.* (2011) have developed high-throughput screening and *in silico* data transformation tools that can speed up *in vitro* hazard ranking. Their multiparametric, automated screening assay incorporates sublethal and lethal cellular injury responses to perform high-throughput analysis of commercial metal/metal oxide nanoparticles with the inclusion of a quantum dot (QD1). Their high volume data set was used for *in vitro* hazard ranking as well as showing the similarity patterns of NPs and response parameters. The *in vitro* results were then compared with the *in vivo* response outcomes in zebrafish embryos. The impact of specific nanoparticles on mortality rate, hatching rate, cardiac rate, and morphological defects was assessed. They have proposed the use of a high-throughput screening, *in silico* data handling and zebrafish testing as a paradigm for rapid and integrated engineered nanomaterial toxicological screening.

The same research group had previously developed a multiparameter, rapid cytotoxicity test that evaluated toxic oxidative stress to compare the effects of titanium dioxide ($TiO_2$), cerium oxide ($CeO_2$), and zinc oxide (ZnO) nanoparticles in bronchial epithelial and macrophage cell lines (George *et al.*, 2010). The nanoparticles were chosen on the basis of their volume of production and likelihood of spread to the environment. Their study demonstrated the utility of a rapid throughput, integrated biological oxidative stress response pathway to perform hazard ranking of selected metal oxide nanoparticles, in addition to showing how this assay can be used to improve nanosafety.

The same group has also developed combinatorial libraries of engineered nanomaterials that permit examination of specific precisely controlled properties to obtain a better understanding of the mechanism at the nano–bio interface (Thomas *et al.*, 2011). While they have accomplished this for a small group of nanomaterials, they propose to extend this methodology to create a compositional library of nanomaterials belonging to different classes that will aid in a better understanding of quantitative structure–activity relationships.

## Exposure Assessment

### The choice of an appropriate exposure metric

Exposure, broadly speaking, is defined as the intensity (i.e., concentration) of the contaminant, at an appropriate interface between the environment and the individual, over a specific time interval that has biological relevance (e.g., an adverse health outcome). There is uncertainty regarding the appropriate metric by which intensity should be measured for inhalation and dermal exposures (ISO, 2007; NIOSH, 2006).

The choice of a more appropriate exposure metric will be critical to determine an exposure–response relationship. For occupational exposure assessment, workers are assigned to similar exposure groups (SEGs). Each SEG represents a group of workers believed to have the same general exposure profile (Ignacio & Bullock, 2006). Designating SEGs appropriately on the basis of the correct exposure information is important for future occupational epidemiologic studies that will investigate the relationship between aerosol contaminants and health effects. If exposure concentration is determined by a less relevant exposure metric when assigning workers to categories, then workers could be misclassified into incorrect categories resulting in a less exposure–response association (Flegal *et al.*, 1991).

Ever since a good correlation between pneumoconiosis and particle mass concentration was determined (Chamberlain *et al.*, 1970), mass concentration has been used as the most appropriate exposure metric for studying associations of particle exposures with ill health (except for fiber exposures). Occupational exposure limits for particle exposure were established on the basis of mass concentration. However, the appropriateness of the mass concentration metric for nanoparticles has been called into question because nanoparticles feature high particle counts and large surface area per mass. While mass concentration has traditionally been used as the metric for exposure assessment of airborne particles and the basis for regulation, it may not always be appropriate for nanomaterials (McCawley *et al.*, 2001). Exposure assessments that rely on mass concentration could underestimate ultrafine particle toxicity, since these particles do not contribute significantly to total mass concentration despite their high numbers. Kreyling *et al.* (2006) reported that the proportion of nano-sized particles is less than 10% of PM2.5 concentrations in terms of mass but more than 90% of the fine particle number concentration. Thus, a case could be made that mass-based occupational exposure limits for nanoparticles should be re-evaluated and other factors such as number or surface should be considered seriously.

Serita *et al.* (1999) exposed rats to different concentrations of ultrafine nickel particles and found that only a single exposure of ultrafine nickel particles caused severe lung damage at the Japanese occupational exposure limit (1 mg/m$^3$) established on the basis of fine nickel particle data. Thus, for the purpose of controlling or managing nanoparticle exposures, mass-based exposure limits may not be useful or effective.

A change of the exposure paradigm for nano-sized particles from a mass basis has been suggested (Kreyling *et al.*, 2006; Maynard & Aitken, 2007). Several studies have espoused particle number concentration as an appropriate metric (McCawley *et al.*, 2001; Peters *et al.*, 1997), while others have championed surface area concentration (Brown *et al.*, 2001; Maynard, 2003; Oberdörster, 2000; Tran *et al.*, 2000). Despite such studies that have reported better dose–response relationships with surface area versus mass concentrations, the toxicological mechanisms of nanoparticles regarding surface area dose have not yet been understood completely.

Additionally, a tremendous amount of uncertainty arises due to changes in physical and chemical properties that can occur when bulk materials with known properties are manufactured at the nano-scale (Fairbrother & Fairbrother, 2009). Nanomaterials can behave in novel and unpredictable ways, challenging researchers to find an understanding of the parameters that contribute to and help predict these properties. The following section highlights several areas of uncertainty for nanomaterials that must be better understood before risks can be assessed.

## Expert and professional judgment in exposure assessment

During the life cycle of a product, different populations will be exposed to the product containing nanomaterials, and their exposure levels and routes can be very different. Exposures need to be assessed for each of these populations in the multiple settings under which exposures could occur—workers, having different occupational tasks, involved in making the product; professionals involved in handling the product during marketing and post-production; consumers in a variety of non-occupational scenarios; and finally workers involved in recycling and disposal.

Usually exposure assessment proceeds from three basic methodologies: subjective estimation using professional judgment, (b) direct measurement of the environment exposure, and (c) prediction of exposure through mathematical modeling. Exposure assessment is conducted on a more routine basis only in occupational settings. Even in these settings, subjective judgments made with little transparency have traditionally driven most exposure assessments, while direct measurements have played a secondary role. Most epidemiological studies relating exposure to chemicals with health effects have been conducted in occupational settings. If exposure assessment needs to be conducted for the general population for a wide range of exposure scenarios, then it will most likely be a combination of exposure modeling and subjective professional judgment and will not involve actual monitoring. The experience of the REACH regulations in the European Union (REACH, 2006) is instructive in this regard. REACH requires that all companies producing or importing substances in quantities of 10 tons or more annually will have to conduct a chemical safety assessment and document this in a Chemical Safety Report. The assessment must include an exposure assessment and risk characterization, and the burden is on the supplier to demonstrate that the risks are adequately controlled in all the various contexts (exposure scenarios) in which the substance may be used or handled. Since many of the possible uses of each chemical will have no monitoring data, physicochemical exposure modeling is the most feasible method for assessing exposures in the thousands of exposure scenarios. However, in the absence of a legal requirement, subjective exposure judgments will also play a major role.

Occupational hygienists utilize skills that can be considered a mixture of science and art to anticipate, recognize, evaluate, and control workplace hazards. An ultimate goal for any exposure assessment would be to accurately define the exposure distribution for a single person or group of people over the time frame of interest. A well-known strategy for exposure assessment and management can be found in the second and third editions of AIHA's Exposure Assessment and Management Book (Ignacio & Bullock, 2006). The strategy recommends that all exposures be characterized qualitatively initially and, where necessary, refined quantitatively by collecting samples that estimate worker exposure. Exposure assessors review the workforce, materials, exposure agents, tasks, equipment, and exposure controls and identify exposure groups that will be assessed and possibly controlled depending on the final judgments. The exposure judgment for any SEG requires the selection of an OEL and a judgment by the hygienist about where the decision statistic (e.g., the 95th percentile of the exposure distribution for the SEG) falls in relation to the OEL. The AIHA strategy illustrates a method for defining control categories or exposure ratings around the OEL for each exposure agent (Table 5.2).

**Table 5.2** Exposure category rating scheme. An SEG is assigned an exposure rating by comparing the 95th percentile exposure ($X_{0.95}$) of the exposure distribution with the full-shift, TWA occupational exposure limit, OEL or STEL (short-term exposure limit), to determine which category it most likely falls into.

| AIHA exposure rating | Proposed control zone description | General description | AIHA recommended statistical interpretation |
|---|---|---|---|
| 1 | Highly controlled (HC) | 95th Percentile of exposures rarely exceeds 10% of the limit. | $X_{0.95} \leq 0.10$ OEL |
| 2 | Well controlled (WC) | 95th Percentile of exposures rarely exceeds 50% of the limit. | $0.10$ OEL $< X_{0.95} \leq 0.5$ OEL |
| 3 | Controlled (C) | 95th Percentile of exposures rarely exceeds the limit. | $0.5$ OEL $< X_{0.95} \leq$ OEL |
| 4 | Poorly controlled (PC) | 95th Percentile of exposures exceeds the limit. | OEL $< X_{0.95}$ |

*Source:* Ignacio and Bullock (Copyright AIHA, 2006)

The exposure assessor assigns an initial "exposure rating," which represents the category where the decision statistic most likely falls; this is based on the information available at the time. Most commonly, the assessor uses a combination of professional judgment, personal experience with a given type of operation, review of exposures or monitoring data from similar operations, and/or exposure predictions developed using physical/chemical exposure modeling techniques to place SEGs into one of the control categories. Experience in making exposure judgments is likely to be related to accuracy and certainty in exposure judgments, probably much like other experiences in life.

Acceptability is commonly evaluated by comparing the true group 95th percentile to the OEL, and on the basis of this comparison, the exposure is classified into one of four categories: "highly controlled," "well controlled," "controlled," or "poorly controlled." A judgment can be documented for each SEG, which can represent a single task that may be short in duration or may represent a group of tasks that comprise a full-shift exposure. Qualitative and quantitative exposure assessments are performed after a thorough review of available information and data related to the workforce, jobs, materials, worker interviews, exposure agents, exposure limits, work practices, engineering controls, and protective equipment.

A decision is represented by a chart showing the assessor's assessment of the probabilities that the 95th percentile lies in each of the four categories (Fig. 5.1).

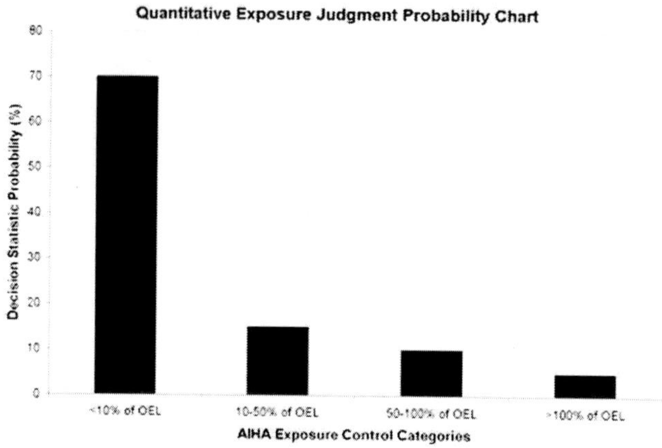


**Figure 5.1** Example qualitative exposure judgment chart illustrating an occupational hygienist's exposure judgment given the information and data available. This chart shows that the hygienist is highly confident the 95th percentile falls into Category 1: <10% of the OEL.

However, a difficulty with this or any exposure management strategy comes when hygienists are required to make exposure judgments with very limited information and data. Any strategy where occupational hygienists make exposure judgments without adequate information or data has the potential to introduce inaccuracy and bias, which could leave people unprotected. The ambiguous process of making exposure judgments with inadequate information has sometimes been referred to as the art of professional judgment.

A number of studies have reported on the accuracy of such subjective exposure judgments in the occupational context, and these findings may be relevant to the general consumer context as well (Logan *et al.*, 2009, 2011; Vadali *et al.*, 2012a; 2012b). The studies involved both desktop assessments (where participating exposure assessors viewed videos of tasks, task information, and sampling data) and walkthrough assessments (where they directly observed the task). In the desktop exposure assessments, 77 exposure assessors participated to evaluate eight tasks that had between 8 and 14 exposure monitoring data points associated with them. There were a total of 4386 participant–task–chemical judgments for all data collections: 552 qualitative judgments made without sampling data and 3834 quantitative judgments with sampling data. In the walkthrough assessments, 49 hygienists from 6 companies participated in the study, and 22 tasks were evaluated. The number of monitoring data points for individual tasks varied between 5 and 24. Ninety-three qualitative judgments (i.e., without reviewing monitoring data) and 2142 quantitative judgments (i.e., those made after reviewing monitoring data) were obtained. The hygienists who decided to participate in the study had varying levels of experience, education, and related determinants for a group of common tasks found in industry. The industrial hygienists were asked to review all available information for each task that they evaluated, including the workplace process, chemicals used in the task, material safety data sheets, controls, personal protective equipment worn by the workers, duration of the task, and task environment parameters such as the general room ventilation. The hygienists were also encouraged to personally view the actual facility and talk to the plant personnel regarding the tasks. The hygienists were asked to give their initial judgment on the basis of their knowledge of the task alone, without looking at any monitoring data. The first monitoring data point was then revealed to them, and they were allowed to modify their initial judgment on the basis of this new information. Subsequent data points were revealed, one at a time, and the hygienists were allowed to modify their judgment after each data point. The judgments were investigated to study how well industrial hygienists predict the correct exposure category as well as the deviation of judgments from the correct category. Since there are four exposure categories, a maximum of over-prediction or under-prediction by three categories is possible. Individual judgment (quantitative & qualitative) categories were compared to the reference category to study accuracy (Figs. 5.2 & 5.3). The key findings relating to quantitative judgments (made using monitoring data) are as follows:

- The accuracy of exposure judgments made by hygienists when monitoring data available is low (<50% correct judgments) but still better than random chance (25%).
- There is a significant underestimation bias in the exposure judgments, that is, there is marked tendency to assign a lower exposure category than the correct one, thus increasing occupational risk to workers.
- The low accuracy is likely due to cognitive biases in understanding skewed log-normal distributions. A training focused on heuristics relating to log-normal statistics significantly improves accuracy to ~70%.
- Several factors relating to cumulative professional experience, training, certification, and educational level of the hygienists as well as task-specific experience were significant predictors of judgment accuracy.

The findings related to qualitative judgments (when no monitoring data are available) are as follows:

- The accuracy of exposure judgments made by hygienists when monitoring data not available (30%) is not much different from random chance (25%).
- The underestimation bias is significant in this case as well.

It is this second set of findings relating to qualitative judgment accuracy (Fig. 5.3a,b) that is more problematic, although the quantitative accuracy findings are related as well. The vast majority of the exposure judgments made by practitioners is qualitative and in many cases even determines if any measurements should be made. The low accuracy of these judgments can therefore lead to incorrect follow-up activities, including risk assessment, and is therefore a cause for concern. It illustrates that the understanding of how workplace factors affect exposure needs to be significantly improved among practitioners. The problem will be especially acute for assessing exposure to nanomaterials. Ramachandran *et*

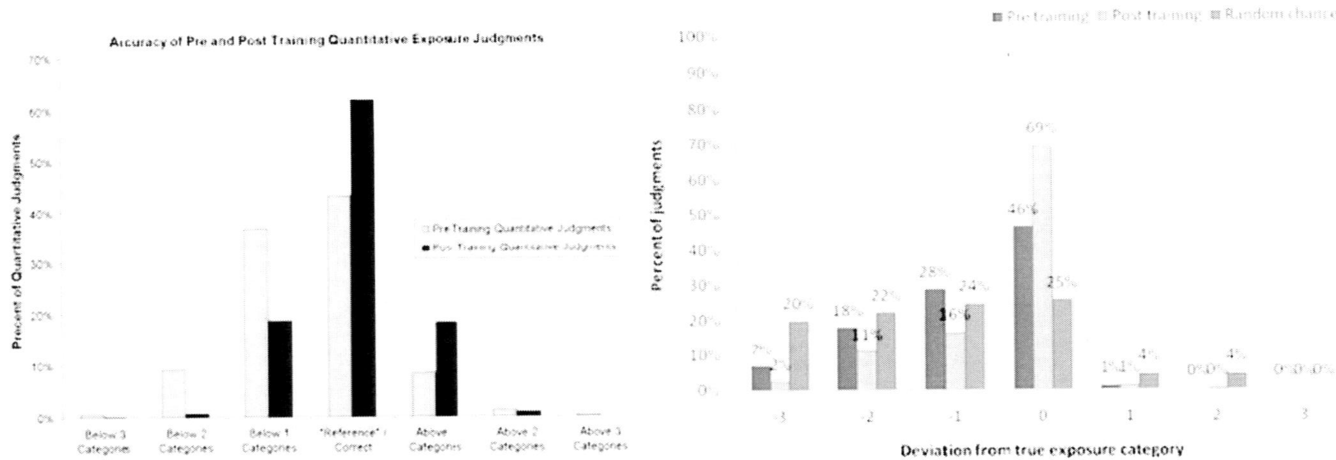


**Figure 5.2** (a) Percentage of all pre- and post-training quantitative task judgments above, below and reference categories for (a) desktop study, $N$ = 3834, and (b) walkthrough, $N$ = 2142.

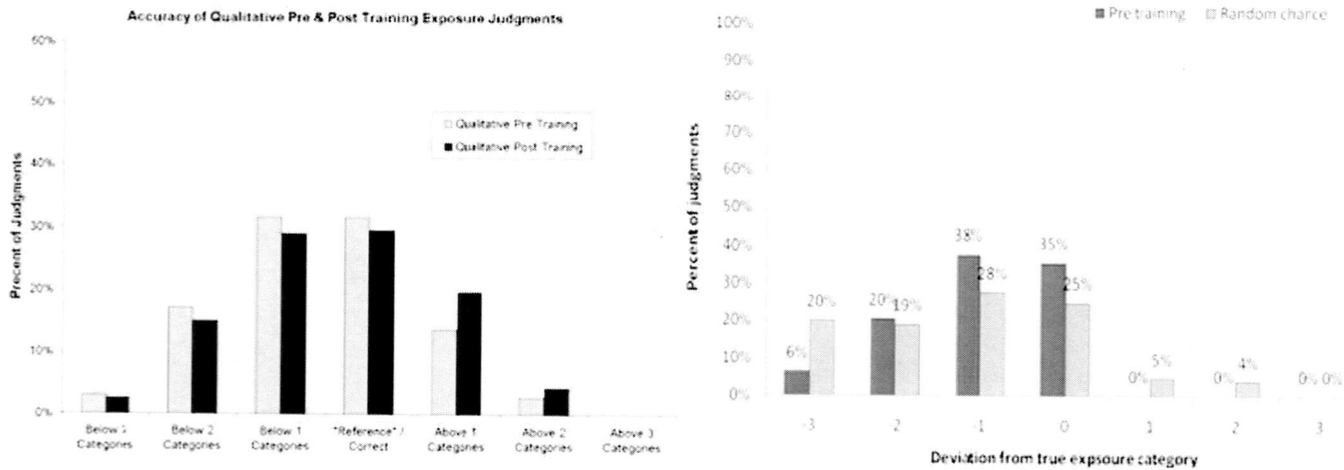


**Figure 5.3** (a) Percentage of all pre- and post-training qualitative task judgments above, below and reference categories for (a) desktop study, $N = 552$, and (b) workplace walkthrough study, $N = 93$.

*al.* (2011b) describe how even subtle differences in nanomaterial characteristics may potentially change their exposure potential, and, thus, the previous experiences of the industrial hygienists may not be applicable and may cause biases leading to erroneous assessments. For example, relying on an obvious visible dust source to recognize the potential for exposures may not be an appropriate indicator for nanomaterial exposure. Likewise, the higher diffusivity of aerosol nanomaterials as well as their interactions with other aerosol particles can often cause unexpected variations in assessing the temporal and spatial distribution of aerosol nanomaterials.

### Heuristics and biases in judgment

The biases in exposure assessments arrived at through subjective means are a result of the use of heuristics by the practitioners. The current literature in psychology relating to human judgment with uncertainty proposes that a limited number of simplifying heuristics are used to efficiently arrive at a judgment using available information. These heuristics or mental processes do not typically utilize all of the available information and data in a formal algorithmic process but use more of "fast and frugal" rules of thumb to arrive at a judgment (Gigerenzer & Todd, 1999; Gilovich *et al.*, 2002; Kahneman *et al.*, 1982). Heuristics can be thought of as simple, efficient rules, hard-coded by evolutionary processes that have been proposed to explain how people make decisions, come to judgments, and solve problems, typically when facing complex problems or incomplete information. The 2002 Nobel Prize in Economics was awarded to Kahneman and Tversky for their work in Prospect Theory, which defined several heuristics in economics related to purchasing decisions (Kahneman & Tversky, 1979). More recently, the theory of heuristics has taken on a more encompassing scope that includes both intuitive and reflective or conscious decision-making systems (Gigerenzer & Todd, 1999; Gilovich *et al.*, 2002).

Studies on various heuristics and their common biases cover a wide range of fields, including psychology, economics, medicine, and engineering. They indicate that humans use simple heuristics rather than extensive algorithms when making judgments (Gigerenzer & Todd, 1999; Kahneman *et al.*, 1982). The three heuristics include: (1) anchoring and adjustment, (2) availability, and (3) representativeness. These simple and efficient cognitive processes in some cases may be biased because of flaws in their simple algorithms

(Kahneman *et al.*, 1982). The amount of bias related to each of these heuristics is usually related to the certainty of the judgment and availability of supporting information and data. Early on, Kahneman and Tversky concluded that people fail to anticipate regression to the mean, fail to give adequate weight to sample size in assessing the import of evidence, fail to take full advantage of base rates when making predictions and possibly other unknown cognitive biases. Their three heuristics were offered as an explanation as to why some errors may occur in a variety of professions (Gilovich *et al.*, 2002).

The anchoring and adjustment heuristic may be used when judgments are made using a single data point or piece of information to make a judgment. People will tend to use the single data point or anchor and make adjustments from it to arrive at the final judgment. An example is a person who is looking to buy a car using only the odometer reading and year to make a final judgment on a used car purchase without evaluating maintenance records, accident history, and test drive. This purchase judgment would be efficient, requiring only two data points, however, would likely be biased and inaccurate (Kahneman *et al.*, 1982). The human brain has a very fast and powerful memory that serves as the architecture for the availability heuristic. Judgments always require some utilization of memory processing and can be driven by what information is available and most salient in memory. It is normal and quite correct in many cases to assume commonality with things that appear similar. When a person is asked to judge the probability that object A belongs to class B, or that event A originates from process B, they will typically rely on what is called the "representativeness heuristic." With this heuristic, people tend to evaluate probability by the degree to which A is representative of, or resembles, B. One possible bias is an insensitivity to sample size, where people may expect a small sample to be representative of the parent population without recognizing that small samples are subject to greater variability. This is the basis for the representativeness heuristic, which uses memory and cognitive features together to form judgments quickly and efficiently (Gigerenzer & Todd, 1999; Griffin & Tversky; Kahneman & Tversky, 1979). The "anchoring and adjustment heuristic" refers to the tendency when making an estimate to start from an initial value (an anchor) and adjust that value up or down. The initial value may be suggested by the formulation of a problem or may be inferred based on a partial calculation. Unfortunately, people tend to stick closely to their initial value and make insufficient adjustments (Kahneman *et al.*, 1982; O'Hagan *et al.*, 2006). While this strategy enables people to reduce computational effort or processing time when performing the same judgment repeatedly, it can result in judgment bias.

To illustrate these three heuristics using a hypothetical example for exposure assessment, consider a hygienist that has four years experience working in the chemical industry and is asked to estimate the upper-tail-exposure distribution for workers handling a new material containing small amounts of methylene chloride on a new process using a special filter press. The hygienist may use the *availability* heuristic to recall from memory all past exposure assessments on a similar filter press using a different material containing xylene. The hygienist may also utilize the *representativeness* heuristic and may pore over the basic characterization information to identify the similarity of this new material, chemicals, and process with the recalled exposure assessments to determine if they likely match. In addition, the hygienist may use the *anchoring and adjustment* heuristics to take information recalled from a representative process exposure assessment and adjust the upper-tail judgment higher on the basis of higher vapor pressure of methylene chloride than xylene. The most likely case would be for the hygienist to utilize a combination of these and other heuristics to arrive at the final judgment.

Kahneman and Tversky's three popular heuristics first proposed are probably only a small subset of other heuristics that exist commonly. In addition, they are not each used exclusively but may be used together and with other unknown heuristics in different combinations (Griffin & Tversky, 1992). Studies indicate that it is not possible to determine a priori which heuristic governs a given judgment even when a limited number of heuristics are investigated (Gilovich *et al.*, 2002). Consider the example where people may use different heuristics when asked whether more people die from rattlesnake bites or from bee stings. If a person had recent heard a news report about snake bites in a given state, they may utilize the availability heuristic by using the relative availability of the two choices when making the judgment. Alternatively, a person may not have any instances coming to mind and might consider the relative "dangerousness" of snake bites and bee stings, which would be an example of the representativeness heuristic (Gilovich *et al.*, 2002).

Some studies indicate that heuristics used in human judgment originate from different sources: Some are hard-coded through the evolutionary process and some can be obtained through conscious learning. A study of Chinese 4th, 5th, and 6th graders and MBA students showed that the level of education strongly influenced the ability to solve Bayesian problems. The authors indicated that one likely reason for the finding was that educational training instilled heuristics that helped solve Bayesian problems (Zhu & Gigerenzer, 2006). This study and others showed that people may be more successful making judgments using frequency formats rather than probability formats without additional training or education. It does appear that training and education helps to close the gaps in judgment effectiveness between frequency and probability formats (Gigerenzer & Todd, 1999). Several authors have concluded that humans likely use a form of dual processing consisting of intuitive heuristics and rational or reason-based heuristics. The intuitive heuristics may rely more on "feeling" or emotions, while the rational heuristics clearly use cognitive rules of thumb or simple algorithms.

Several studies indicate that overconfidence is very common and propose several possible determinants of certainty (Griffin & Tversky, 1992). They suggest that certainty of judgments is based on both the strength of evidence and the credibility or weight of the evidence. The authors propose that judgment overconfidence is determined by the strength and balance of arguments for and against the competing hypotheses, without proper regard for the weight of the evidence such as sample size. The authors give many examples of overconfidence in judgment making with non-experts and also professionals across several fields. In addition, the authors present study results that show how overconfidence can be caused when the sample parameter values are high and the sample size is low. They also show in the same study that underconfidence was created when the sample parameter values were low and the sample size was high.

The above findings and discussion of biases illustrate the potential and risks of using expert judgments as a risk assessment tool. The above example related to the ability of experts and professionals to assess personal exposures in occupational settings. However, expert judgments are used in a wide array of situations within the umbrella of risk assessment. Expert judgment has been used in many contexts when uncertainty is high and is well suited for the challenges posed by emerging nanomaterials (Kandlikar *et al.*, 2007). The key is to be aware of the possibility of the judgments being biased and guarding against this. Studies such as those reported in the previous section suggest some techniques to achieve this—training to prevent underestimation of risk and the use of models instead of direct assessments, that is, the disaggregation of the judgment into several components. In the absence of objective data, the recourse to using expert judgments is widespread in many fields and is justifiable. Given the known problems with using expert judgments, elicitation exercises must account for and correct for these biases.

## Extreme uncertainty in risk assessment models

Traditional risk assessment and decision-making techniques often employ mathematical models and subjective expert judgment when uncertainty is high (Fryer *et al.*, 2006). The reliance on models, be they formally mathematical or qualitative, is a property of all prediction. Mathematical models used in risk analysis inherently contain various degrees of epistemic uncertainty. These arise due to two reasons: the world is inherently stochastic and many quantities of interest can only be represented by probability distributions; secondly, models are incomplete by necessity, and in the domains that are incorporated in the model, they may be inaccurate. Uncertainties in values of model parameters play a central role in conventional risk calculations (*parametric uncertainty*), as does uncertainty about the proper form of models (*structural or model uncertainty*). Disagreements may also occur between scientific studies or expert's subjective judgments, thereby increasing uncertainty over what would constitute "proper" or "accurate" models or variables.

Uncertainties in the *values of parameters* in risk assessment models often arise due to measurement errors that are inherently stochastic (e.g., errors in measuring emissions from a reactor producing nanomaterials), systematic biases (e.g., using an area monitor for to estimate exposures for workers who are much closer to the nanomaterial source), *natural variation* (e.g., due to changes in systems with respect to time, space, or other variables in ways that are difficult to predict), and non-representativeness (e.g., estimating a exposure–response curve from an *in vitro* study conducted with a sample that does not represent the general population). In contrast, *model uncertainty* arises when the relationships within and among various components of the risk assessment framework are poorly understood. There may be several competing models to explain relationships between variables of interest, or in the worst case, no models may exist at all. Therefore, the disconnect between reality and a model lies not in the choice of values for the parameters of a model, but in the choice of model itself. While parametric uncertainty represents the known unknown and we know the structure of the missing information, model uncertainties are inherently difficult to express within a probabilistic framework. Model gaps represent lack of awareness of a relevant part of reality, that is, unknown unknowns, and by necessity it is impossible to ascribe probability distributions to quantities or relationships that the modeler is unaware of.

### *"Fat tail" distributions*

For handling parametric uncertainty, probabilistic methods such as Monte Carlo analysis have traditionally been used to generate estimates of expected values for the outcomes of interest. Risk models often describe the relationship between exposures and health endpoints or surrogates of health endpoints (e.g., a biomarker of effect or expression of a protein). For many chemical exposures, the exposure distribution is found to be log-normal in nature. And indeed when exposure is modeled using a physicochemical model and Monte Carlo simulations are run, the resulting histogram is right-skewed and a log-normal distribution can usually be fit to the histogram. This is true even when we assign various types of distributions to the input parameters. However, when the input parameter distributions are assigned small variabilities, the resultant output distribution resembles a normal rather than a skewed distribution. This has serious consequences for risk assessment and management. Assumptions of normality for various input parameters or a derived output distribution are problematic, since probabilities decrease rapidly as we move away from the mean (Taleb, 2010). Thus, large variations away from the mean are deemed to have a very low probability of occurrence. This is a recipe for underestimating risk.

This was illustrated spectacularly in the economic collapse of 2008 that upset many common assumptions about the global financial system. Several expert commentators have pointed out that severe economic crises, in general, are not rare events. Although these large crises happen every decade on average, standard risk models do not assign a significant probability that these events will occur. Traditional methods of modeling risk often fail to reflect the frequency of declines and when these declines will occur. Such models use an assumption of normality and use the standard deviation as a measure of risk. Models based on this assumption predict that a return more than three standard deviations below average should have occurred only one or two times since 1926. However, the stock market's history shows these extreme negative events have occurred 5 to 10 times more than predicted by the models (e.g., eight peak-to-trough declines in the S&P 500 of over 20% since the mid-1920s).

The problem is even more acute for describing an unknown unknown, that is, ascribing probability distributions to model uncertainties. One suggestion has been the use of meta-probabilities—that is, ascribing probabilities or degrees of belief in each probability. In other words, a probability distribution has an associated metaprobability distribution that describes the probability of the probability distribution being correct. This has been described by Taleb (2007) by putting error rates on the Gaussian distribution. Taleb (2007) describes the variance of the distribution as epistemologically equivalent to the lack of knowledge about the expected value and the variance of the variance as being equivalent to lack of knowledge about the lack of knowledge. The variance of the variance is analogous to the fourth moment of the distribution, that is, its kurtosis. Thus, the presence of heavy (or fat) tails is analogous to the lack of knowledge about the lack of knowledge.

There is an alternative distributional approach that assigns non-trivial probabilities to large percentage changes. This model assigns probability to its tails in a manner akin to the "power law," so that as the magnitude of loss increases, the probability of loss decreases gradually, giving the distribution fat tails. True fat-tailed distributions have a scale-free or fractal property. While this more accurately reflects stock market returns, it is not widely used due to mathematical complexity.

Another significant fallacy in thinking about the risk of extreme events is that even if several experts have significant differences in initial belief, their updated beliefs arising from available data would converge on the same final belief using a process called Bayesian updating. However, this convergence would work only if the experts agreed on the underlying structure of the data. If the data are for rare events, then multiple distributions could fit the same data set. In such a situation, different expert priors combined with different likelihoods of observing the data would lead to different posterior beliefs.

## Risk management using control bands

There have been several practical approaches to decision-making in limited realms. Control banding is a methodology that has been used extensively by the pharmaceutical industry for categorizing exposure controls in the workplace (Zalk & Nelson, 2008). The idea is to develop "control bands" that can be mapped to particular sets of exposure/hazard combinations, so that health risks for workers involved in particular operational tasks are minimized. Each control band corresponds to a particular control technology or action that is suitable for the given hazard and exposure scenario.

A nanomaterial-specific control banding methodology (CB Nanotool) was developed by Paik *et al.* through extensive expert input, review, and testing (Paik *et al.*, 2008; Zalk *et al.*, 2009). The CB Nanotool involves a basic matrix with "severity" (i.e., hazard) and "probability" (i.e., exposure) indices as the two axes and utilizes nanomaterial physical/chemical properties (shape, size, surface area, & surface activity), available toxicology information (carcinogenicity, mutagenicity), and exposure availability (volume produced, dustiness) to link the indices to one of the four control bands (Paik *et al.*, 2008). The control bands correspond with increasingly stringent control methodologies, from "general ventilation" up to "seek specialist advice." An example of the CB Nanotool control banding matrix can be seen in Fig. 5.4.

| Severity Score \ Probability Score | Extremely Unlikely (0 - 25) | Less Likely (26 - 50) | Likely (51 - 75) | Probable (76 - 100) |
|---|---|---|---|---|
| Very High (76 - 100) | RL 3 | RL 3 | RL 4 | RL 4 |
| High (51 - 75) | RL 2 | RL 2 | RL 3 | RL 4 |
| Medium (26 - 50) | RL 1 | RL 1 | RL 2 | RL 3 |
| Low (0 - 25) | RL 1 | RL 1 | RL 1 | RL 2 |

**Figure 5.4** Control banding matrix with risk-level (RL) indicators as a function of the combination of probability and severity scores. Control bands correspond to risk levels as follows: RL 1—General Ventilation; RL 2—Fume hoods or local exhaust ventilation; RL 3—Containment; RL 4—Seek specialist advice. Adapted from Paik *et al.* (copyright 2008, Oxford University Press).

Analysts can input known "severity" factors for each nanomaterial in use and estimate "probability" factors based on specific exposure characteristics of each occupational task under consideration. The resulting combination of severity and exposure scores for each task relate to one of the four control bands that provide control advice for minimizing occupational risk. One of the strengths of such a tool lies in its ability to utilize basic data without the need for expensive testing.

### *Robust versus optimal decision-making for risk management*

Traditional decision sciences have revolved around identifying an optimal decision, given various constraints and criteria. An optimal decision is one that scores the most desirable values on the various criteria of interest. Multi-criteria decision analysis (MCDA) is a widely discussed approach to nanomaterial risk assessment with a long history of use in various decision contexts (Linkov *et al.*, 2007; Tervonen *et al.*, 2009). The control banding approach described in the previous section can be considered an example of such a schema. While MCDA is a useful tool, the simple modeling techniques that underlie it (linear, additive response models) can miss the actual complexity of the risk phenomena. Mechanism-based models that characterize the relevant physical and biological variables and their interactions provide more accurate representations and so are more scientifically defensible. Additionally, it is not clear whether the MCDA framework is well suited to analysis under high uncertainty.

The following is an example of extreme model uncertainty: consider the risk assessment for an automobile. If we take the automobile to be only an ensemble of various chemicals and conduct a chemical risk assessment, then parametric uncertainties would include quantities like the amounts of various heavy metals and plastics used in the production, the exposures of workers to these chemicals during production and the health risks from such exposures, the emissions of the automobile during operation, and the risks to the general population from automobile exhaust components, among other things. However, this model does not account for the risks arising from the fact that an automobile is not just an ensemble of chemicals; it has a specific function, namely locomotion, that has its own set of risks of accidents and injuries, risk factors such as the age of the driver, the road conditions, the age of the vehicle, and the wear on the tire treads. Not including such factors or not being aware of them is an example of a model gap that can cause serious risk underestimation.

The above analogy is apt for nanomaterials. Risk assessment for nanomaterials has thus far been considered only as a variation on chemical risk assessment. However, with the development of smart nanomaterials and functional nanomachines, it would be erroneous to consider any nanomaterial as only an assembly of chemicals, albeit with some unusual properties. At the same time, as nanomaterials are combined in nanotechnologies and nanosystems, the assessment of risks is more than just a summation of the risks of the individual components. The complex interactions will need to be considered. Lifeycle analysis will also face these challenges as the devices and nanosystems containing these nanomaterials later decompose and have to be disposed of in a responsible manner.

Thus, the requirement of protecting against gaps in our knowledge and parametric uncertainties requires a shift in perspective—from seeking an optimal decision to seeking a robust solution. Given several alternatives, a robust decision is one whose outcomes are less sensitive to model assumptions and uncertainties. Such a decision would be less susceptible to the effects of surprise. A new area of decision sciences called robust adaptive planning (Bankes, 2002; Lempert *et al.*, 2002) focuses on studying the performance of decisions under a large ensemble of plausible future scenarios. A decision is considered robust is it performs reasonable well over this wide range of scenarios, compared to the alternatives.

There are several dimensions to robustness. A simple type of robustness refers to the sensitivity of the output to variability in one or several input parameters. Here, a lack of sensitivity is an indication of robustness. Conversely, if the output is indeed sensitive to the input, then courses of action that may reduce this sensitivity can be determined. A second type of robustness can be tested by identifying assumptions whose validity is necessary for the decision to be correct and leading to the desired outcome. The third type of robustness testing is to explore the existence of model gaps. However, these strategies are heavily dependent on human heuristics and their associated biases that have been discussed in an earlier section.

Another aspect of robustness is worth mentioning. Redundancy is a positive attribute in risk-based decision-making. Functional redundancy is where the same function can be performed by several structures within the system (i.e., availability of spare parts). If the system of risk management has redundancy built into it, then it allows for the system to survive adversity or surprise outcomes. Robust systems are more resilient than optimized systems that have low degrees of redundancy. An optimal system does not have any functional redundancy. Another type of redundancy is where a structure within a system can be used to perform a function that is not its central one and that it was not specifically designed for. This is a type of adaptation that facilitates opportunistic decision-making (Taleb, 2010).

## Conclusions

The overarching theme in this chapter is the nature and sources of uncertainties in risk assessment in the context of life cycle analysis. The three main sources of uncertainties are (1) insufficient objective knowledge of the mechanisms of action of nanomaterials and the appropriate metric of human exposure, (2) the limitations and biases in expert assessment of risk and its components, and (3) the limitations and biases in the use of risk models.

As the idea of life cycle risk assessment for nanomaterials takes root, it is important for practitioners to address some key issues—roadblocks as well as opportunities—that affect more widespread use of the concepts embedded in such assessments. While life cycle analysis incorporates materials and energy use as well as human and ecological health effects, this chapter has focused more on the latter. Within the realm of health risk assessment, nanomaterials are characterized by great uncertainties with regard to toxicity and exposure assessments. While novel nanomaterials may have physical and chemical properties that may affect standard *in vitro* toxicity tests, systems biology offers unprecedented opportunities to understand the mechanism of effects by understanding the complex signaling that occurs in biological organisms. While choice of appropriate exposure metrics that are health-relevant remains uncertain, sophisticated instrumentation is available, which allow us to measure all of the candidate metrics with great precision. These are tools with great promise for the purpose of assessing risks from novel technologies.

Another key idea presented here is that comprehensive life cycle analysis requires consideration of cognitive and behavioral factors.

There has been a great deal of quantitative work in recent years that has described the biases in exposure judgments made by professional occupational hygienists and the role of various heuristics in human decision-making. These biases tend to underestimate exposure and hence risk. However, targeted training appears to remove the bias to some extent. Finally, increased attention has focused in recent years on the nature of extreme risks, the implications for modeling and statistical analysis, and robust decision-making under extreme uncertainty. These again underline the importance of explicit consideration of cognitive factors in life cycle analysis, especially for novel technologies with great uncertainties. Risk itself is a way of framing the future impacts of nanomaterials and technologies that assume that experts can provide an assessment of "real risks" as an output. This assumption is valid only when the inputs to the risk assessment process (i.e., objective data regarding our scientific understanding of the exposure & toxicity issues) and the transfer function (i.e., the structure & assumptions of risk models as well as cognitive structures that process these inputs) are unbiased. A clear understanding of these interactions is important for progress in life cycle risk assessment.

## References

Ashford, N. A., & Caldart, C. C (1996). Technology, law and the working environment. Washington, DC: Island Press.

Balbus, J. M., Florini, K., Denison, R. A., *et al.* (2007). Protecting workers and the environment: An environmental NGO's perspective on nanotechnology, *Journal of Nanoparticle Research*, 9, 11–22.

Bankes, S.C. (2002). Tools and techniques for developing policies for complex and uncertain systems, *Proceedings of the National Academy of Sciences*, 99(3), 7263–7266.

Becker, R. A., Borgert, C. J., Webb, S., *et al.* (2006). Report of an ISRTP workshop: Progress and barriers to incorporating alternative toxicological methods in the U.S., *Regulatory Toxicology and Pharmacology*, 46, 18–22.

Bowman, D., & Hodge, G. (2007). Nanotechnology and public interest dialogue: Some international observations, *Bulletin of Science Technology & Society*, 27, 118–132.

Brown, D. M., Wilson, M. R., MacNee, W., *et al.* (2001). Size-dependent proinflammatory effects of ultrafine polystyrene particles: A role for SA and oxidative stress in the enhanced activity of ultrafines, *Toxicology and Applied Pharmacology*, 175, 191–199.

Chamberlain, E. A. C., Makower, A. D., & Walton, W. H. (1970). *New Gravimetric Dust Standards and Sampling Procedures for British Coal Mines. Inhaled Particles III* (pp. 111–124). London, UK: Unwin Brothers Ltd.

Chatterjee, R. (2008a). The continuing uncertainty of nano risks and regulations, *Environmental Science & Technology*, 42(21), 7733.

Chatterjee, R. The challenge of regulating nanomaterials, *Environmental Science & Technology*, 42(2), pp. 339–343.

Choi, J., Ramachandran, G., & Kandlikar, M. (2009). The impact of toxicity testing costs on nanomaterial regulation, *Environmental Science & Technology*, 43(9), 3030–3034.

Donaldson, K., Borm, P. J., Oberdorster, G., *et al.* (2008). Concordance between *in vitro* and *in vivo* dosimetry in the proinflammatory effects of low-toxicity, low-solubility particles: The key role of the proximal alveolar region, *Inhalation Toxicology*, 20, 53–62.

Fairbrother, A., & Fairbrother, J. (2009). Are environmental regulations keeping up with innovation? A case study of the nanotechnology industry, *Ecotoxicology and Environmental Safety*, 72(5), 1327–1330.

Flegal, K. M., Keyl, P. M., & Nieto, F. J. (1991). Differential misclassification arising from nondifferential errors in exposure measurement, *American Journal of Epidemiology*, 134, 1233–1241.

Fryer, M., Collins, C., Ferrier, H., *et al.* (2006). Human exposure modeling for chemical risk assessment: A review of current approaches and research and policy implications, *Environmental Science & Policy*, 9(3), 261–274.

George, S., Pokhrel, S., Xia, T., *et al.* (2010). Use of a rapid cytotoxicity screening approach to engineer a safer zinc oxide nanoparticle through iron doping, *ACS Nano*, 4, 15–29.

George, S., Xia, T., Rallo, R., *et al.* (2011). Use of a high-throughput screening approach coupled with *in vivo* zebrafish embryo screening to develop hazard ranking for engineered nanomaterials, *ACS Nano*, 5(3), 1805–1817.

Gigerenzer, G., & Todd, P. M. (1999). *ABC Research Group. Simple Heuristics that Make Us Smart*. New York, NY: Oxford University Press.

Gilovich, T., Griffin, D., & Kahneman, D. (2002). *Heuristics and Biases. The Psychology of Intuitive Judgment*. New York, NY: Cambridge University Press.

Griffin, D., & Tversky, A. (1992). The weighing of evidence and the determinants of confidence, *Cognitive Psychology*, 24, 411–435.

Haimes, Y. Y., & Lambert, J. H. (1999). When and how can you specify a probability distribution when you don't know much? II, *Risk Analysis*, 19(1), 43–46.

Henry, C. J., Phillips, R., Carpanini, F., *et al.* (2002). Use of genomics in toxicology and epidemiology: Findings and recommendations of a workshop, *Environmental Health Perspectives*, 110, 1047–1050.

Ignacio, J., & Bullock, B. (eds.) (2006). *A Strategy for Assessing and Managing Occupational Exposures* (3rd ed.). Fairfax, VA: AIHA Press.

International Organization for Standardization (ISO) (2007). *Workplace Atmospheres—Ultrafine, Nanoparticle and Nanostructured Aerosols—Inhalation Exposure Characterization and Assessment*, ISO/TR 27628: 2007(E). Geneva: ISO.

Jeong, S.-H., Cho,W.-S., Kim, J.-E., *et al.* (2011). Systems toxicological approach to the risk assessment of nanomaterials, *General, Applied and Systems Toxicology*. Doi: 10.1002/9780470744307.gat247.

Kahneman, D., Slovic, P., & Tversky, A. (1982). *Judgment Under Uncertainty: Heuristics and Biases*. New York, NY: Cambridge University Press.

Kahneman, D., & Tversky, A. (1982). Prospect Theory: An analysis of decision under risk, *Econometrica*, 47, 263–291.

Kandlikar, M., Ramachandran, G., Maynard, A., *et al.* (2007). Health risk assessment for nanoparticles: A case for using expert judgment, *Journal of Nanoparticle Research*, 9, 137–156.

Klöpffer, W., Curran, M. A., Frankl, P., *et al.* (2007). *Nanotechnology and Life Cycle Assessment.* Washington, DC, 2–3 October, 2006. Synthesis of Results Obtained at a Workshop, Project on Emerging Nanotechnologies. Washington, DC: Woodrow Wilson International Center for Scholars.

Kohler, A. R., Som, C., Helland, A., *et al.* (2008). Studying the potential release of carbon nanotubes throughout the application life cycle, *Journal of Cleaner* Production, 16, 927–937.

Kreyling, W. G., Semmer-Behnke, M., & Möller W. (2006). Health implications of nanoparticles, *Journal of Nanoparticle Research*, 8, 543–562.

Kroll, A., Pillukat, M. H., Hahn, D., *et al.* (2009). Current *in vitro* methods in nanoparticle risk assessment: Limitations and challenges, *European Journal of Pharmaceutics and Biopharmaceutics*, 72, 370–377.

Kuempel, E., Geraci, C., & Schulte, P. (2007). Risk assessment approaches and research needs for nanomaterials: An examination of data and information from current studies, Nanotechnology – Toxicological *Issues and Environmental Safety and Environmental Safety*, 5, 119–145.

Lane, N., & Kalil, T. (2005). The national nanotechnology initiative: present at the creation. *Issues Sci. Technol.*, Summer 2005. Available at http://www.issues.org/21.4/lane.html (accessed on April 5, 2010).

Lave, L. B., Cobas-Flores, E., Hendrickson, C. T., *et al.* Using Input-output analysis to estimate economy-wide discharges, *Environmental Science & Technology*, 29(9), 420A–426A.

Lempert, R., Popper, S., & Bankes, S. (2002). Confronting surprise, *Social Science Computer Review*, 20(4), 420–440.

Linkov, I., Satterstrom, F., Steevens, J., *et al.* (2007). Multi-criteria decision analysis and environmental risk assessment for nanomaterials, *Journal of Nanoparticle Research*, 9, 543–554.

Lloyd, S. M., & Lave, L. B. (2003). Life cycle economic and environmental implications of using nanocomposites in automobiles, *Environmental Science & Technology*, 37, 3458–3466.

Lloyd, S. M., Lave, L. B., & Matthews, H. S. (2005). Life cycle benefits of using nanotechnology to stabilize platinum-group metal particles in automotive catalysts, *Environmental Science & Technology*, 39, 1384–1392.

Logan, P. W., Ramachandran, G., Mulhausen, J. R., *et al.* (2011). Desktop study of occupational exposure judgments: Does education and experience influence accuracy? *Journal of Occupational and Environmental Hygiene*, 8, 746–758.

Logan, P. W., Ramachandran, G., Mulhausen, J.R., *et al.* (2009). Occupational exposure decisions: Can limited data interpretation training help improve accuracy? *Annals of Occupational Hygiene*, 53(4), 311–324.

Lux Research. *The Nanotech Report* (5th ed). New York, NY: Lux Research Inc.

Lyndon, M. L. (1989). Information economics and chemical toxicity: Designing laws to produce and use data, *Michigan Law Review*, 87, 1795–1861.

Maynard, A. D. (2003). Estimating aerosol surface area from number and mass concentration measurements, *Annals of Occupational Hygiene*, 47, 123.

Maynard, A. D., & Aitken, R. J. (2007). Assessing exposure to airborne nanomaterials: Current abilities and future requirements, *Nanotoxicology*, 1, 26–41.

McCawley, M. A., Kent, M. S., & Berakis, M. T. (2001). Ultrafine beryllium number concentration as a possible metric for chronic beryllium disease risk, *Applied Occupational and Environmental Hygiene*, 16, 631–638.

National Institute for Occupational Safety and Health (NIOSH). Approaches to Safe Nanotechnology: An Information Exchange with NIOSH. Version 1.1, 2006.

Nel, A., Xia, T., Madler, L., *et al.* (2006). Toxic potential of materials at the nanolevel, *Science*, 311, 622–627.

Oberdörster, G. (2000). Toxicology of ultrafine particles: *In vivo* studies, *Philosophical Transactions of the Royal Society A*, 358, 2719–2740.

O'Hagan, A., Buck, C., Daneshkhah, A., *et al.* (2006). *Uncertain Judgements: Eliciting Experts' Probabilities.* West Sussex, England: John Wiley & Sons.

Paik, S., Zalk, D., & Swuste, P. (2008). Application of a pilot control banding tool for risk level assessment and control of nanoparticle exposures, *Annals of Occupational Hygiene*, 52, 419.

Peters, A., Wichmann, H. E., Tuch, T., *et al.* (1997). Respiratory effects are associated with the number of ultrafine particles, *The American Journal of Respiratory and Critical Care Medicine*, 155, 1376–1383.

Ramachandran, G., Ostraat, M., Evans, D. E., *et al.* (2011a). A strategy for assessing workplace exposures to nanomaterials, *Journal of Occupational and Environmental Hygiene*, 8, 673–685.

Ramachandran, G., Wolf, S. M., Paradise, J., *et al.* (2011b). Recommendations for oversight of nanobiotechnology: Dynamic oversight for complex and convergent technology, *Journal of Nanoparticle Research*, 13, 1345–1371.

REACH. Regulation (EC) No. 1907/2006 of the European Parliament and of the Council of 18 December 2006 concerning the Registration, Evaluation, Authorisation and Restriction of Chemicals (REACH), establishing a European Chemicals Agency, amending Directive 1999/45/EC and repealing Council Regulation (EEC) No 793/93 and Commission Regulation (EC) No 1488/94 as well as Council Directive 76/769/EEC and Commission Directives 91/155/EEC, 93/67/EEC, 93/105/EC and 2000/21/EC. *Official Journal of the European Union*, EN, L396, 1–841.

Roco, M. C. (2003). Broader societal issues of nanotechnology, *Journal of Nanoparticle Research*, 5(3–4), 181–189.

Roco, M. C. (2007). Nanotechnology initiative: Past, present, future. In W. A. Goddard *et al.* (Eds.), *Handbook on Nanoscience, Engineering, Technology* (2nd ed., pp. 19–44). Boca Raton, FL: Taylor & Francis.

Sayes, C. M., Reed, K. L., & Warheit, D. B. (2007). Assessing toxicity of fine and nanoparticles: Comparing in vitro measurements to in vivo pulmonary toxicity profiles, Toxicological sciences: An official journal of the Society of Toxicology, 97, 163–180.

Serita F., Kyono H., & Seki Y. (1999). Pulmonary clearance and lesions in rats after a single inhalation of ultrafine metallic nickel at dose levels comparable to the threshold limit value. *Industrial Health*, 37, 353–363.

Taleb, N. N. (2007). *Dynamic Hedging: Managing Vanilla and Exotic Options*. New York, NY: John Wiley & Sons.

Taleb, N. N. (2010). *The Black Swan: The Impact of the Highly Improbable*. New York: Random House Inc.

Tervonen, T., Linkov, I., Figueira, J., *et al.* (2009). Risk-based classification system of nanomaterials, *Journal of Nanoparticle Research*, 11, 757–766.

Thomas, C. R., George, S., Horst, A. M., *et al.* (2011). Nanomaterials in the environment: From materials to high-throughput screening to organisms, *ACS Nano*, 5(1), 13–20.

Tran, C. L., Buchanan, D., Cullen, R. T., *et al.* (2000). Inhalation of poorly soluble particles. II. Influence of particle surface area on inflammation and clearance, *Inhalation Toxicology*, 12, 1113–1126.

United States General Accounting Office (US GAO). (2005). *Options Exist to Improve EPA's Ability to Assess Health Risks and Manage its Chemical Review Program, (GAO-05-458)*. Washington, DC: U. S. Government Printing Office. Available at http://www.gao.gov/new.items/d05458.pdf.

United States General Accounting Office (US GAO). (1994). *Toxic Substances Control Act: Legislative Changes Could Make the Act More Effective (GAO/RCED-94-103)*. Washington, DC: U. S. Government Printing Office. Available at http://archive.gao.gov/t2pbat2/152799.pdf.

Vadali, M., Ramachandran, G., & Banerjee, S. (2012a). Effect of training, education, professional experience and need for cognition on decision making in occupational exposure assessment, *Annals of Occupational Hygiene*, 56(3), 292–304.

Vadali, M., Ramachandran, G., Mulhausen, J. R., *et al.* (2012b). Effect of training on exposure judgment accuracy and agreement among hygienists, *Journal of Occupational and Environmental Hygiene*, 9(4), 242–256.

Wagner, W. E. Common ignorance: the failure of environmental law to produce needed information on health and the environment, *Duke Law J*. 2004, 53, pp. 1619–1745.

Zalk, D., & Nelson, D. (2008). History and evolution of control banding: A review, *Journal of Occupational and Environmental Hygiene*, 5, 330–346.

Zalk, D., Paik, S., & Swuste, P. (2009). Evaluating the control banding nanotool: A qualitative risk assessment method for controlling nanoparticle exposures, *Journal of Nanoparticle Research*, 11, 1685–1704.

Zhu, L., & Gigerenzer, G. (2006). Children can solve Bayesian problems: Tthe role of representation in mental computation, *Cognition*, 98, 287–308.

## Chapter 6

# Organizational Capability Life Cycles for Responsible Innovation

**Paul Ellwood,[a] Krsto Pandza,[a] and Erik Fisher[b]**
[a]*Leeds University Business School, University of Leeds, Leeds, LS2 9JT, UK*
[b]*Center for Nanotechnology in Society at Arizona State University, Arizona State University, PO Box 874401, Tempe, AZ 85282-4401, USA*
p.ellwood@leeds.ac.uk

## Introduction

The study of innovation has proved of substantial interest to scholars in management and organizational studies (MOS) with work ranging from the macro level (economy/industry/market) through organizational level (firm/enterprise) to the micro level (group/team/individual). In addition to the substantial amount of research on the organization of innovation (Crossan & Apaydin, 2010), management practitioners and policy makers continue making substantial investments in this area in order to increase business competitiveness (Kao, 2009). Our contribution to this book's exploration of life cycle perspectives is to adopt an organizational level of analysis in order to present the process dynamics of

*Emerging Technologies: Socio-Behavioral Life Cycle Approaches*
Edited by Nora Savage, Michael Gorman, and Anita Street

ISBN 978-981-4411-00-4 (Hardcover), 978-981-4411-01-1 (eBook)
www.panstanford.com

innovation. We are explicitly interested in how organizations whose purposes are served by the development of new technologies are influenced by life cycle considerations.

In seeking to address this question, we will explore an organizational-level life cycle perspective based on the concept of *capabilities* and illustrate this discussion with a contemporary context by relating our interest in the management of emerging technology with the policy discourse on responsible innovation (RI). Organizational capabilities are a well-established construct in the field of strategic management (Dosi *et al.*, 2000) and have been defined as collectively held and action-oriented knowledge that enables organizations to get things done (Pandza & Thorpe, 2010). The management of emerging technology in order to develop new products would be an example of an organizational capability (Eisenhardt & Martin, 2000). Because of the fluid nature of new technology markets, any capability associated with their applications is likely to be ineffective if it remains stationary and unresponsive to the changing pressures of the business environment. Therefore, it becomes important to understand how organizational capabilities might evolve through their own life cycles (Helfat & Peteraf, 2003). However recognizing that such a life cycle process might be subjected simultaneously to different influences, we find it instructive to decompose such cycles into more basic process theories of change (Van de Ven & Poole, 1995); not least to facilitate their empirical study.

The first section opens with a core issue for emerging technologies; that of how the inherent uncertainty in their reception and impact might be managed. This theme is given a particular resonance with the introduction of a contemporary context for the management of emerging technologies in terms of the policy discourse on responsible innovation (RI). Organizational capabilities for innovation and their evolution will be presented in the hope of offering fresh insights on how the aspiration of RI policy discourse may be met by innovative enterprises. On the basis of our research, ideas will be proffered on the micro-foundational practices that constitute an emerging organizational capability for RI. Finally a typology of process theories of change is introduced as a means of achieving a more granular perspective on the progression of organizational capability life cycles.

## Emerging Technologies and Responsible Innovation

In this section, we outline the different threads of research and policy work that is being conducted under the heading of responsible innovation. It is our contention that an organizational level of analysis has been relatively underdeveloped in this policy discourse; and yet if innovating enterprises are to be brought into this envisioned arena, then this will need to change, and practical strategies for their engagement will need to be developed.

For organizations who seek to launch new science-based products, the well-documented *uncertainty* of discontinuous technological change (Anderson & Tushman, 1990) is compounded by a fragile and often difficult-to-understand public sentiment toward an emergent technology. Social studies of public perceptions of nanotechnology (Currall, 2009; Currall *et al.*, 2006; Scheufele *et al.*, 2007) suggest a complex and nuanced set of factors that can influence social acceptance of the emergent technology. This intrinsic uncertainty that accompanies technology dynamics presents a central challenge for regulatory frameworks and risk management techniques. In growing recognition of the need for a more nuanced and shared governance approach to innovation, multi-stakeholder engagement with technologies has taken on new significance to a variety of actors. Understanding multi-stakeholder engagement requires us to go well beyond the understanding of innovation as commercially viable technological invention with a potential impact on competitive dynamics (Christensen & Rosenbloom, 1995; Tushman & Anderson, 1986)—the view that predominantly populates the management literature.

A contemporary science policy discourse that has been labeled *responsible innovation* reinforces the suggestion that core business practices in general, and technology innovation practices in particular, can no longer afford to view innovation as a direct route toward acquiring wealth; the notion of innovating *responsibly* suggests shifting attention from very tangible economic and financial benefits to more open-ended concepts, such as adding value to society (Enderle, 2009) or, as Von Schomberg (2011) suggests, co-creating value with society. Whilst no one definition of RI can claim wide acceptance, a recent suggestion by Von Schomberg

(2011) is reproduced here as it offers a helpful introduction to the challenges under debate: "Responsible Research and Innovation is a transparent, interactive process by which societal actors and innovators become mutually responsive to each other with a view to the (ethical) acceptability, sustainability, and societal desirability of the innovation process and its marketable products (in order to allow a proper embedding of scientific and technological advances in our society)."*

Discourses surrounding RI have corresponded to a number of policy enactments (European Commission, 2005; US Congress, 2003), and different dimensions to this policy work may be discerned. Understanding and implementing these enactments pose numerous conceptual, practical, and cultural challenges. For instance, heightened awareness of the importance of attending to the time lags (Owen & Goldberg, 2010), complex relationships (Stokes, 1996), and early uncertainties (Scheufele *et al.*, 2007) entailed over the course of conducting original research, developing new products, and seeking to promote the societal embedment of science-based technology has resulted in more openness on the part of some innovation actors to calls for more adaptive and anticipatory forms of governance in the conduct of innovation (Karinen & Guston, 2010). Furthermore, understanding the complexity of the many issues associated with any particular technology-enabled product requires drawing from varied perspectives of risk management (Shatkin & North, 2010), policy studies (Sabatier, 1999), innovation studies (Schuurbiers, 2011), science and technology studies (Selin, 2011), and other forms of scholarship. Meanwhile, pressures to engage with a wide range of stakeholders on the challenges posed by emerging technologies like nanotechnology have resulted in large organized public dialogues and debates (Cobb, 2011; Jones, 2009; Stilgoe, 2007), but how these feed back into innovation processes is not clear.

As complements to these large-scale public engagement exercises, which can include research and innovation actors and policy actors as intended audiences for the reports they generate, recent forms of Technology Assessment (TA) and Ethical, Legal, and

---

*We note that this vision of "responsible innovation," which touches on the manner in which innovation is conducted, differs in kind from corporate social responsibility (CSR). CSR is largely a matter of contributions out of profits to socially beneficial causes that are not central to business and technology innovation plans and activities of the firm or organization in question.

Social Implications (ELSI) research seek to integrate humanities and social scientific methods and result into research and innovation processes. Unlike these newer instantiations, earlier forms of TA and ELSI research were meant to provide information primarily to policy makers about ongoing research authorization or emerging regulatory challenges. In order to represent and contrast these different approaches and to specify the potential links between them and the science and innovation actors who also need to be included in schemes for RI, we offer a rubric to conceptualize disparate efforts over time and across actor groups that might be aimed at RI (see Fig. 6.1). As implied by Fig. 6.1, one can differentiate the governance of science and innovation into three overlapping and mutually constitutive stages. Although this is a simplification, it highlights something that is novel about RI discourses and goals. For, not only are "upstream and downstream" policies making stages important to include in efforts to conceptualize and implement RI, but the "midstream" stage of science and innovation policy implementation should also be included.

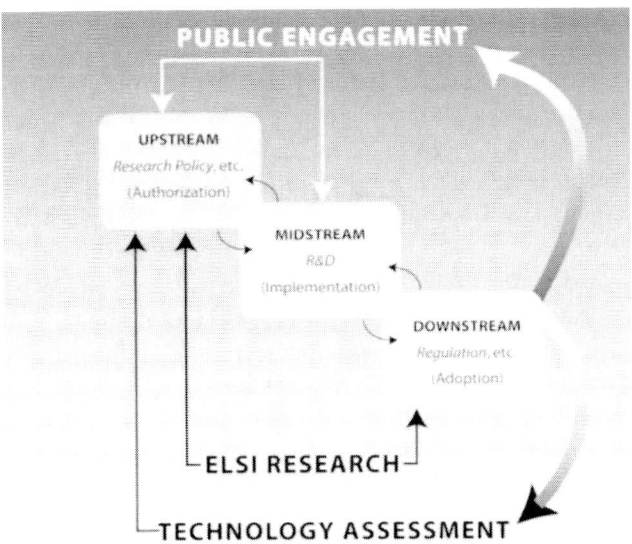


**Figure 6.1** Science and technology governance stages. Reproduced with permission from Fisher *et al.* (copyright 2006, SAGE Publications).

Formal approaches to technology assessment that attempt to take into account these and other related challenges and that have gained traction with science policy makers include "Constructive Technology Assessment" (Rip *et al.*, 1995) and "Real-Time Technology Assessment" (Guston & Sarewitz, 2002). Both CTA and RTTA depart from earlier forms of technology assessment that were meant to provide "early warning" to policy makers about future regulatory challenges. Adopting a more distributed view of innovation, CTA and RTTA aim to interact with and inform technology and innovation actors and processes in order to productively and responsibly broaden them. They both include efforts aimed at "midstream modulation" (Fisher *et al.*, 2006), namely, efforts that are informed by a view of science and innovation actors as governance agents. CTA and RTTA efforts are intended to interact practically and over time with the midstream. For instance, RTTA combines multiple methods and attempts to take multiple sites and stages of research, policy, and innovation into account—in particular, anticipation (Selin, 2011), public engagement (Cobb, 2011), and socio-technical integration (Fisher, 2007) activities (cf. Barben *et al.*, 2008).

These and similar midstream or multi-stream TA approaches both inform and are informed by elements of the discourse on RI. This discourse, in turn, arguably plays a perceptible role in the development of regulation and in the public funding of research and innovation with nanotechnologies. And yet in spite of these evolving policy discourses, enactments, and efforts at public and practitioner engagement, it remains unclear as to how to conceptualize and practically interact with innovating organizations in modes that extend beyond that of simply responding as they have to prior regulations. Our interest is to explore how RI might be built into the organizational strategies and practices of innovating enterprises so that they might help contribute to the promise of the policy discourse (Von Schomberg, 2011). While some CTA and RTTA workshops have involved a range of innovation partners, including industry (Fisher *et al.*, 2010; Rip, 2005), more generally, efforts to connect innovating enterprises to this agenda have been limited. As policy instruments are developed to enact RI, greater consideration will need to be given to the involvement of such innovation partners, and particularly those from private industry. In this regard, recent initiatives that have sought to engage university research groups in this agenda through funding incentives (Owen & Goldberg, 2010) and by engendering socio-ethical reflexivity (Fisher, 2007) are potential candidates for transfer to private industry.

It is noteworthy that much of the literature dealing with the topic of RI is found in the domain of science studies rather than management and organization studies. In the remainder of this chapter, we seek to supplement organization studies literature with this science studies literature so that we might better understand the less-examined meso-level implications of the policy discourse on RI. Such policy discourse implies a fundamental shift in the operating environment of innovating organizations; ones whose impact goes beyond the conventional considerations of changes in the macro-environment, as they throw into question issues of corporate purpose, vision and identity, rather than simply having implications for competitive strategy. Strategic responses of the type a firm may deploy to a change in the macro-environment (e.g., organizational structural change, product line extensions, new technology platform, etc.) will, we argue, be insufficient as a response to a requirement to innovate responsibly. In order to explore a meso-level (e.g., a business-level) evolution in innovation practice consistent with these emerging RI mandates, we will make use of the concept of *organizational capabilities*.

## Organizational Capabilities for Innovation

Organizational capabilities are conceptualized as collectively held and action-oriented knowledge that enables organizations to get things done (Pandza & Thorpe, 2010). They are often thought of as being embedded in habitual patterns of organizational learning (Zollo & Winter, 2002). Through collective learning organizations develop capabilities that enable them to grow in competitive environment. The dynamic of this learning, however, depends on changes in markets. If changes in markets are frequent, then large organizations face constant pressure to recombine resource and renew capabilities.

Within a business context, "a capability for getting things done" often means achieving a competitive advantage over market rivals. Management and organization scholars have long been interested to explain how firms secure and maintain such advantages in their chosen markets. One traditional argument is that competitive advantage is dependent on industry structure and how an individual firm aligns itself in relation to that structure (Porter, 1979). More recently, another influential school of thought, known as the "resource-based view of the firm" (RBV), has maintained that competitive advantage results from the distinctive internal organization of resources within the firm itself (Barney, 1991). Firms operating within the arena of emerging technologies present a particular challenge to the RBV perspective because of the inherent dynamic nature of the markets in which they operate (Teece *et al.*, 1997). The competitive landscape of such markets is in a constant state of flux given the inherent uncertainties associated with emerging technologies, their disruptive nature, and practical efficacy. As a result, the *dynamic capabilities* by which managers "integrate, build, and reconfigure internal and external competencies to address rapidly changing environments" (Teece *et al.*, 1997, at 516) become the key to achieving competitive advantage.

Helfat and Peteraf extended this conception of dynamic capabilities to include *all* organizational capabilities (dynamic or static) by theorizing on how all capabilities can be understood to evolve through life cycles (Helfat & Peteraf, 2003). Evoking life cycle thinking in this manner allows for a more complete understanding of the evolution of how an organization might marshal its resources in order to successfully manage emerging technologies. Such resources and the new products they create are considered to be "opposite sides of the same coin" (Wernerfelt, 1984). It is suggested that there is a clear relationship between the life cycles through which those products pass and the life cycles of the organizational capabilities used to generate them. However, this relationship is not a direct one-to-one correspondence (Helfat & Peteraf, 2003) as the two types of life cycles are likely to operate across different timelines. An organization might reasonably have a portfolio of products each at different stages in their life cycle, while the corresponding capability life cycle is necessarily at a single point at any given time. It follows that a major shift in an organizational life cycle might be expected to have implications for not only the new products that an organization has in development but those in more advanced growth or maturity stages in their life.

The different stages in a capability life cycle suggested by Helfat and Peteraf (2003) are shown in Fig. 6.2. The horizontal axis is a proxy for time and represents the cumulative amount of activity associated with the capability. The vertical axis is intended to show the level of the capability employed per unit of activity. A move up the vertical axis would represent an increase in the degree of competence at the given capability. The diagram is generic in nature and illustrates a number of different evolution scenarios. The branching of the life cycle and different evolutionary trajectories will be influenced by a number of factors: the antecedents in the capability evolution up to that point, strategic change decisions taken internally by the organization's managers, and the impact of changes in the external environment of the organization.

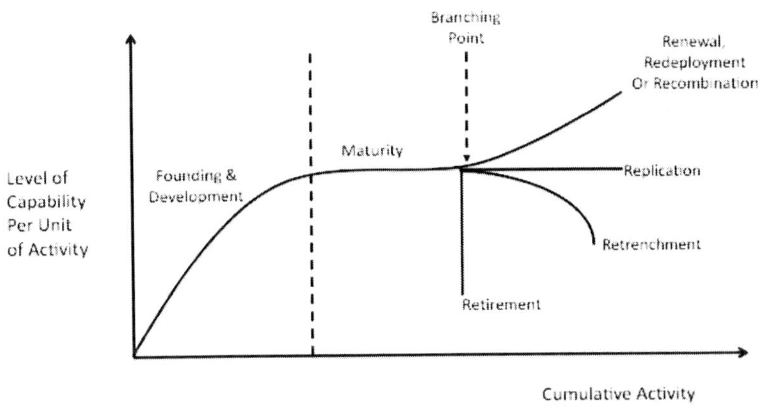


**Figure 6.2** Stages of the organizational capability life cycle. Adapted from Helfat and Peteraf (copyright 2003, John Wiley & Sons, Inc).

Helfat and Peteraf explain the progression between the different stages in the capability life cycle by imagining the development of a new organization created to achieve a specific objective that requires the evolution of a new capability. The *Founding* stage is built upon the natural endowments of the newly created organization (including, crucially, the skills of the people within it). The *Development* of these initial conditions follows a classical organizational learning-by-doing trajectory (Argyris & Schon, 1996). This stage is characterized by activities such as the search for alternative approaches and trial-and-error experiments, and it is difficult to identify a clear transition point between the *Founding* and the *Development* stages. The level of capability that the organization is able to create will eventually plateau, or attain *Maturity*, as the limit of what is possible with the available resources is reached.

In the absence of a provocation for change, the Maturity stage in the capability life cycle might continue indefinitely, with the only challenge being one of maintaining the level of the capability. Branching points (see Fig. 6.2) might be triggered by factors that are internal or external to the organization, and such factors could be either threats or opportunities. At such a juncture, Helfat and Peteraf (2003) outline six generic strategies for capability transformation that alter the shape of the subsequent life cycle trajectory, and which they term "the six Rs:"

- *Retirement*, or withdrawal of the capability
- *Retrenchment*, or the gradual decline in the level of the capability
- *Replication*, or copying the capability in some other context (e.g., a different geographical market for the same product)
- *Redeployment*, or applying the capability to a different product
- *Recombination*, or adding the original capability to another
- *Renewal*, or improvement of the capability in some way

For management and organizational scholars, theories which account for the evolution of resources and capabilities are essential for understanding the core challenge of how an organization might achieve a *sustainable* competitive advantage in its chosen field of endeavor. Helfat and Peteraf's model for capability life cycles widens our perspective on organizing-for-innovation over time beyond that of more well-known product life cycle considerations (Kotler, 1980). In doing so, they advocate (Helfat & Peteraf, 2003) studies in the evolution of organizational capabilities as a separate empirical undertaking.

On the basis of our own such researches, we suggest that the aspiration inherent in the discourse on RI may constitute an emerging branching point for the organizational capability to more reflexively manage emerging technologies. In the next section, we draw upon our experience of engaging with scientists at nanotechnology research centers in order to trace a possible life cycle trajectory for an organizational capability in RI.

## Evolution of an Organizational Capability for Responsible Innovation

The opportunity to study the evolution of an organizational capability for RI arose through the authors' collaboration on the NSF *Socio-Technical Integration Research* (STIR) project. Conceived by Erik Fisher and David Guston of Arizona State University, the project was a response to national government policy pronouncements that science research ought to be more responsive to wider societal concerns. In an extension to established social science lab study methodologies (Hess, 2001), STIR employs an innovative research philosophy that probes the capacity of scientists and engineers to participate in socio-technical integration (Fisher, 2007). The objectives of the STIR project were as follows:

- To identify and compare external expectations and demands for laboratories to engage in RI.
- To assess and compare the current responsiveness of laboratory practices to those pressures.
- To investigate and compare how interdisciplinary collaborations may assist in educating, enhancing, and stimulating responsiveness.

To undertake this research program, PI Fisher built a network of senior collaborators (including co-author Pandza) and PhD students (including co-author Ellwood) from a variety of disciplines (science and technology studies, philosophy, policy science, anthropology, political science, and management research), who have research interests in broadly the same area. In addition to the primary goal of testing and developing a new lab engagement research methodology—midstream modulation (Fisher *et al.*, 2006)—the different disciplinary interests within the STIR network of researchers allowed for other theoretical perspectives to be explored. Within the context of this overarching project, inductive examination of case study data suggested an evolving organizational capability for RI within the subset of two nanotechnology research groups studied. These different elements are presented in Fig. 6.3 in relation to Helfat and Peteraf's (2003) capability life cycle model.

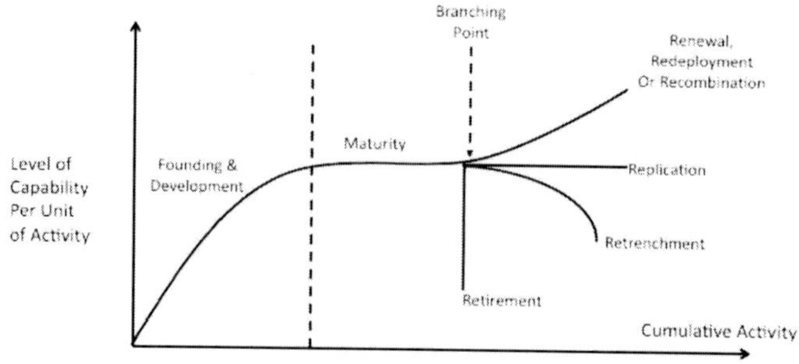


| Stage-relevant Foundational Elements of Responsible Innovation | Pursuit of high quality science<br><br>Collaborations with science partners | Search for new opportunities<br><br>Engagement with wider stakeholders |
|---|---|---|

**Figure 6.3** Evolution of micro-foundational elements of an organizational capability for responsible innovation. Adapted from Helfat and Peteraf (copyright 2003, John Wiley & Sons, Inc).

Given that experienced scientists populate these research organizations, it is to be expected that the conduct of research that is judged by peers to be of high quality would be a foundational element in a capability for science-led innovation. Any scientific idea will be roundly tested through social processes of skepticism and comparison to the established standards of multiple areas of technical and material practices. Interestingly, the research leader of one nanotech research group equates the highest technical standards with being able to make a practical contribution: "to be truly useful you've got to make sure that your tools are right up to date, and of the highest quality."

The creation of modern scientific and technological knowledge has been recognized as a complex process (Powell *et al.*, 1996). Breakthroughs occur at different times and within different organizations. The transfer or diffusion of knowledge between these places becomes critical for the management of innovation. Therefore, as research organizations seek to develop their capability for innovation, collaborating with science partners (e.g., major industrial companies) is the principle means by with they realize their innovation aspirations. Core aspects of their strategic activities become geared toward the realization of innovation projects through collaborations with science-based companies.

Providing projects continue to be successful in both research (e.g., the production of peer-reviewed research papers) and innovation (e.g., the realization of industrially relevant knowledge) terms, most university research groups might remain at this (Maturity) stage in their capability life cycle. And yet during the STIR lab engagement, we encountered scientists at all levels for whom the policy discourse on RI, or the STIR protocol (Fisher, 2007) and midstream modulation framework (Fisher *et al.*, 2006), was prompting to rethink their innovation practice. These findings are consistent with those from other studies that used the same method (Schuurbiers, 2011). This leads us to suggest that as discourse on RI extends its reach, a "branching point" could emerge in the capability life cycle; a capability for RI could very well emerge from an existing capability for innovation. What then might be the practices associated with the *renewal* branch of the capability life cycle?

Conventional wisdom holds opportunity-seeking behavior to be common to most contemporary research groups who are considered to be successful. Indeed, research practitioners saw this behavior as a foundational pursuit of high-quality science. However, we also found an enthusiasm to explore new sources of opportunity, ones grounded in real-world challenges. At one research group, three new enterprises had been "spun-out" and current researchers have regular professional and social contact with their counterparts in these start-up companies. An orientation toward the challenge of commercialization was evident in the way new researchers framed their study's questions, and even in the professional development activities (i.e., securing a scholarship to participate in an entrepreneurship program) undertaken by the most senior researcher in the team. The search for new activities is readily understandable in terms of the organization's efforts to sustain and enhance its position compared to research rivals. In addition, interviews with the research leaders often revealed motivations and interests that went beyond scientific and commercial aspirations. One research director spoke frequently of an interest in environmental challenges, and indeed a number of their existing projects are an expression of this interest. Some of these had a clear scientific dimension, but more intriguing were environmental challenges (e.g., passive energy reduction in building) for which he did not yet have a solution in mind.

The majority of external engagement activities undertaken by research groups are concerned with stakeholders with a direct interest in the scientific and technological outputs (e.g., industrial companies). Interviews with the research leaders and extramural activities of some of the researchers themselves provide evidence for an awareness of a wider set of stakeholders for a modern research enterprise. Such awareness has not translated into norms or organizational practices, but it appears to be characterized by a collection of exploratory activities. In one group, a young researcher, acting through personal motivation, secured a scholarship to participate in a dissemination-of-science event for the "general public." This activity required him to produce a poster explaining the rationale of his research project and its findings for a non-scientific audience. In a similar vein, researchers at another group have hosted visits to their facility by schoolchildren with the aim of encouraging the study of Physics at university. Such engagement events are prompted in part, by the rhetoric of his Director that challenges the next generation of scientists to question, "What is the real value of what you are doing? What's your life mission? Is it to keep producing five scientific papers a year?" These are the questions for which answers were not widely evident in the organizational routines of the research groups. However, they could be taken as signs that the rhetoric of public engagement with science (Jones, 2009; Stilgoe, 2007) has inclined some scientists toward engaging more explicitly with publics.

## Process Theories of Change within Organizations

In this final section, we further unpack the notions of evolution of organizational capabilities by examining different process theories of change as this provides insight into how such evolutionary change may be studied empirically. In general terms, organizational change is defined as "a difference in form, quality, or state over time in an organizational entity" (Van de Ven & Poole, 1995). While such change has often been viewed in terms of its outcomes (Tushman & Romanelli, 1985), in this chapter we have based our discussion on ideas of change, which focus on process, as that allows for the identification of the drivers of such processes while before outcomes of the change have been realized (Van de Ven & Poole, 1995).

Van de Ven (2007) has suggested that there are two definitions of process that are used in the study of organizational change: process as a "concept of categories" and process as a "developmental event sequence." In the first of these definitions, organizational or individual action is understood as a category (e.g., decision making) with clear definitions, which can be distinguished between other organizational categories we may wish to study (e.g., operational environment). This approach views process as consisting of a series of input-transformation-output events that can be studied by operationalizing the categories as variables and examining the change in that variable through the event. Such studies are appropriate for an understanding of the antecedents of change. In contrast, the "development event sequence" approach adopts a longitudinal perspective to the activities under study and is appropriate for studying questions of how change occurs within organizations. Here process is described as a "progression of activities," and one can identify various generic types of progression (Van de Ven, 2007).

In order to explain how and why a process is elaborated over time, Poole and Van de Ven (1995) identified four idealized *theories of process*, following an interdisciplinary literature review of the subject. These authors proposed a typology of four fundamentally different theories of process, with each theory having a different cycle of change and a particular "engine." The representation in Fig. 6.4 of these cycles is in the form of a 2 × 2 matrix whose axes signify different levels of analysis and modes of change. Change processes operate at many different levels from the individual organization to groupings of organizations (e.g., firms within an industry sector). The vertical axis in Fig. 6.4 distinguishes between change at the level of a single organizational entity and that of some more aggregated grouping of entities. Such a distinction draws attention to two different perspectives on the study of organizational change. The first concerns the internal historical processes of change operating within a single organization. The second allows the study of the interaction between organizations and processes, such as cooperation and competition. The mode of change axis considers whether the unfolding of change is either *prescribed* or *constructed*. A constructed mode of change involves the creation of new organizational arrangements that represent a discontinuous shift from the past. By contrast, a prescribed mode of change is one that is regulated by some institutional or natural circumstance at the outset of the change.

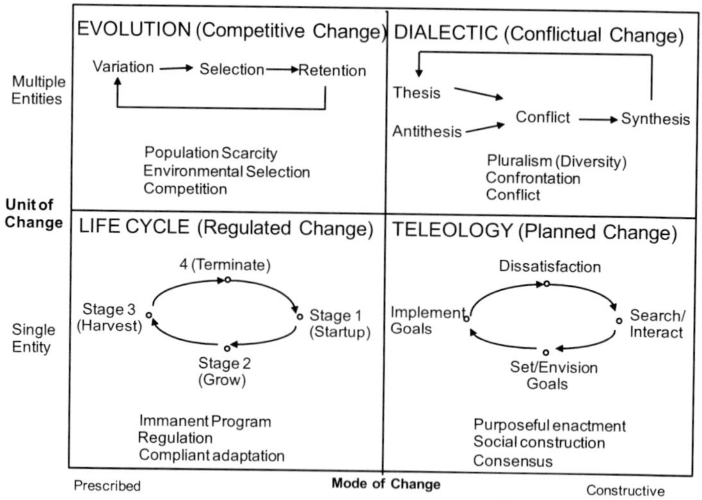


**Figure 6.4** Process models of organization change. Reproduced with permission from Van de Ven and Poole (copyright 1995, Academy of Management).

Considering the bottom-left quadrant, we feel a possible confusion arises in relation to the broader use of the term life cycle in this chapter, in which a specific label of *life cycle* is used to describe models of change that are represented as operating on a single organizational entity, within a prescribed modality. In this manner, change is viewed as imminent; that is to say that the entity undergoing change is considered to have within it an underlying logic that governs the change process. In a sense, the end point of the change is pre-figured in the entity's original conditions (e.g., regulatory mandates). This is not to say that the operating environment of the entity has no influence on the process of change, but rather that such influence is of secondary importance. The unfolding of the change process takes place in a prescribed order in which each stage holds the conditions and logic necessary for the next stage. Also operating at the level of a single entity, but this time involving a more purposeful enactment of change, the bottom-right quadrant illustrates the *teleological* motor. Such change arises from the pursuit of aspirations and speaks to notions of agency within organizations. In this case, the starting point is not fixed and the cycle incorporates learning undergone within the organization as it passes through goal formulation, implementation, review, and modification.

At a more aggregated level of analysis, purposeful change is represented (in top right of Fig. 6.4) as a *dialectical* process, as entities in conflict connect to produce a synthesis that in turn can represent a thesis for the next iteration of the cycle. Such situations can be understood in terms of the relative balance of power or legitimacy between opposing ideas with change resulting from a shift in that balance. The fourth and final model in this typology is termed *evolutionary* by Van de Ven and Poole, as it results from the competition among multiple entities for scarce resources. In an analogous manner to Darwin's natural selection, change here occurs through an on-going process of variation, selection, and retention.

Change in real-world organizations is rarely understandable in terms of a single-process type. In developing their theory, Van de Ven and Poole (1995) suggest that combinations of their four basic engines might be invoked in order to arrive at a better understanding of real-world change. The 16 possible combinations of these 4 basic theories of process are considered in turn, and readings are offered from the literature in order to demonstrate how complex theories of change might be understood in terms of combinations of their four engines. Van de Ven (2007) advocates holding more than one of these theories in mind when undertaking process studies as this allows for a more nuanced interpretation of research data.

It is instructive to unpack our earlier exploration of capability life cycles in terms of these four basic theories of change. Given our interest in the organizational level of analysis, the models in the bottom row of Fig. 6.4 are the ones that we should try to relate to the discussion of capability life cycles. Following the ideas of Helfat and Peteraf (2003), we suggested that the discourse on RI might constitute a branch point in the life cycle of an organizational capability for innovation. The discrete and sudden fashion in which this change event is shown in Figs. 6.2 and 6.3 was always likely to be in incomplete representation of the real-world situation. The case research across the STIR program of lab studies showed both varied degrees of awareness of RI-related policy discourses and different efforts to connect with such discourse (e.g., by trying to reflexively engage with social scientists or forming links with other research stakeholders). Some of the innovation practices we observed during the STIR projects could be read as research groups experimenting with new ways of working. In several instances, attempts to relate their research to broader social contexts and value questions actually improved research practices (Fisher, 2007; Fisher *et al.*, 2010; Schuurbiers, 2011). Thus, whereas the enactment of policy instruments associated with RI might constitute the starting point for the *life cycle* theory of change shown in Fig. 6.4, an adaptive exploration of new innovation practices is better represented by the *teleological* model. A progressive life cycle for an organizational capability for innovation might be understood as being constituted of both prescribed (i.e., responding to policy mandates) and constructed (i.e., aspirational pursuit of new goals) catalysts for change.

## Final Remarks

The life cycle perspective presented in this chapter aims to foreground an organizational level of analysis. In doing so, we have also sought to supplement research in management and organizational studies with insights gleaned from the science studies literature. Taking the contemporary policy discourse on RI as the context for our discussion of emerging technologies, we are interested in understanding the change dynamics of innovative enterprises facing a shifting operating environment, one that requires them to attend to the "(ethical) acceptability, sustainability, and societal desirability of the innovation process and its marketable products" (Von Schomberg, 2011). As attention moves to questions on how private industry might respond to these policy mandates, it is productive to consider how individual innovating organizations might engage with the RI concept. We found a variety of ways in which research practitioners appear to be attempting to translate their own broader social aspirations into what could develop into more established organizational capabilities. These include selection of research projects, public outreach, expanding the range of stakeholders interacted with, and, in some cases, reflection on the social and public value of research during the conduct of that research. Sustained collaborations with social scientists—by not only making new innovation practices more visible but also introducing more critical observations derived from the study of an organization's social context as compared with their own behavior—represent one route toward building capacities for developing new organizational capabilities. Given a focus on the organizations bringing forth emerging technologies, rather than on their technologies and associated products, we have drawn from the management studies literature on organizational capabilities in order to explore innovation dynamics. In this regard, a life cycle framework has allowed us to integrate the responsive (prescribed) and aspirational (constructive) elements that might constitute an organization's strategy for enacting RI.

## Acknowledgment

This material is based on work supported by the National Science Foundation under award number #0849101.

## References

Anderson, P., & Tushman, M. L. (1990). Technological discontinuities and dominant designs—a cyclical model of technological change. *Administrative Science Quarterly*, *35*(4), 604–633.

Argyris, C., & Schon, D. A. (1996). *Organizational Learning II: Theory, Method, and Practice*. Reading, MA: Addison Wesley.

Barben, D., Fisher, E., Selin, C., & Guston, D. H. (2008). Anticipatory governance of nanotechnology: foresight, engagement and integration. In E. J. Hackett, O. Amsterdamska, M. Lynch, and J. Wajcman (Eds.), *The Handbook of Science and Technology Studies* (3rd ed.). Cambridge, MA: MIT Press.

Barney, J. (1991). Firm resources and sustained competitive advantage. *Journal of Management*, *17*(1), 99–120.

Christensen, C. M., & Rosenbloom, R. S. (1995). Explaining the Attackers Advantage—technological paradigms, organizational dynamics, and the value network. *Research Policy*, *24*(2), 233–257.

Cobb, M. D. (2011). Creating informed public opinion: citizen deliberation about nanotechnologies for human enhancements. *Journal of Nanoparticle Research*, *13*(4), 1533–1548.

Crossan, M. M., & Apaydin, M. (2010). A multi-dimensional framework of organizational innovation: a systematic review of the literature. *Journal of Management Studies*, *47*(6), 1154–1191.

Currall, S. C. (2009). New insights into public perceptions. *Nature Nanotechnology*, *4*(2), 79–80.

Currall, S. C., King, E. B., Lane, N., Madera, J., & Turner, S. (2006). What drives public acceptance of nanotechnology? *Nature Nanotechnology*, *1*(3), 153–155.

Dosi, G., Nelson, R. R., & Winter, S. G. (2000). *The Nature and Dynamics of Organisational Capabilities*. Oxford: Oxford University Press.

Eisenhardt, K. M., & Martin, J. A. (2000). Dynamic capabilities: what are they? *Strategic Management Journal, 21*(10–11), 1105–1121.

Enderle, G. (2009). A rich concept of wealth creation—beyond profit maximisation and adding value. *Journal of Business Ethics*, 84, 281–295.

European Commission. (2005). *European Technology Platform on NanoMedicine: Vision Paper and Basis for a Strategic Research Agenda for NanoMedicine*. Luxembourg: Office for Official Publications of the European Communities.

Fisher, E. (2007). Ethnographic invention: probing the capacity of laboratory decisions. *Nanoethics, 1*(2), 155–165.

Fisher, E., Biggs, S., Lindsay, S., & Zhao, J. (2010). Research thrives on integration of natural and social sciences. *Nature, 463*(7284), 1018–1018.

Fisher, E., Mahajan, R. L., & Mitcham, C. (2006). Midstream modulation of technology: governance from within. *Bulletin of Science, Technology and Society, 26*(6), 485–496.

Guston, D. H., & Sarewitz, D. (2002). Real-time technology assessment. *Technology in Society, 34*, 93–109.

Helfat, C. E., & Peteraf, M. A. (2003). The dynamic resource-based view: capability lifecycles. *Strategic Management Journal, 24*(10), 997–1010.

Hess, D. (2001). Ethnography and the development of science and technology studies. In P. Atkinson, A. J. Coffey, S. Delamont, J. Lofland, and L. H. Lofland (Eds.), *Handbook of Ethnography*. London: SAGE.

Jones, R. A. L. (2009). Public engagement and nanotechnology—the UK experience. In J. Stilgoe (Ed.), *Public Dialogue on Science and Technology: The Road Ahead* (pp. 59–69). London: Sciencewise-ERC.

Kao, J. (2009). Tapping the world's innovation hot spots. *Harvard Business Review, 87*(3), 109–114.

Karinen, R., & Guston, D. H. (2010). Towards anticipatory governance: the experience with nanotechnology. In M. Kaiser, M. Kurath, S. Maasen, and C. Rehmann-Sutter (Eds.), *Nanotechnology and the Rise of an Assessment Regime*. Dordrecht, Netherlands: Springer.

Kotler, P. (1980). *Principles of Marketing* (3rd ed.). Englewood Cliffs, NJ: Prentice-Hall.

Owen, R., & Goldberg, N. (2010). Responsible innovation: a pilot study with the UK engineering and physical sciences research council. *Risk Analysis, 30*(11), 1699–1707.

Pandza, K., & Thorpe, R. (2010). Management as design, but what kind of design? an appraisal of the design science analogy for management. *British Journal of Management, 21*(1), 171–186.

Porter, M. E. (1979). How competitive forces shape strategy. *Harvard Business Review, 57*(2), 137–145.

Powell, W. W., Koput, K. W., & Smith-Doerr, L. (1996). Interorganizational collaboration and the locus of innovation: networks of learning in biotechnology. *Administrative Science Quarterly, 41*(1), 116–145.

Rip, A. (2005). *Technology Assessment as Part of the Co-Evolution of Nanotechnology and Society: The Thrust of the TA Program in NanoMed.* Paper presented at the conference on "Nanotechnology in science, economy and society", 13–15 January, Marburg, Germany.

Rip, A., Misa, T. J., & Schot, J. (Eds.). (1995). *Managing Technology in Society: The Approach of Constructive Technology Assessment.* London and New York: Pinter Publishers.

Sabatier, P. A. (Ed.). (1999). *Theories of the Policy Process.* Boulder, CO: Westview Press.

Scheufele, D. A., Corley, E. A., Dunwoody, S., Shih, T.-J., Hillback, E., & Guston, D. H. (2007). Scientists worry about some risks more than the public. *Nature Nanotechnology, 2*(12), 732–734.

Schuurbiers, D. (2011). What happens in the lab: applying midstream modulation to enhance critical reflection in the laboratory. *Science and Engineering Ethics, 17*(4), 769–788.

Selin, C. (2011). Negotiating plausibility: intervening in the future of nanotechnology. *Science and Engineering Ethics, 17*(4), 723–737.

Shatkin, J. A., & North, W. (2010). Perspectives on risks of nanomaterials and nanotechnologies: advancing the science. *Risk Analysis, 30*(11), 1627–1633.

Stilgoe, J. (2007). *Nanodialogues: Experiments in the Public Engagement with Science.* London: Demos.

Stokes, D. E. (1996). *Pasteur's Quadrant: Basic Science and Technological Innovation.* Washington, DC: Brookings Institute.

Teece, D. J., Pisano, G., & Shuen, A. (1997). Dynamic capabilities and strategic management. *Strategic Management Journal, 18*(7), 509–533.

Tushman, M. L., & Anderson, P. (1986). Technological discontinuities and organizational environments. *Administrative Science Quarterly, 31*(3), 439–465.

Tushman, M. L., & Romanelli, E. (1985). Organizational evolution: a metamorphosis model of convergence and reorientation. In B. M. Staw and L. L. Cummings (Eds.), *Research in Organizational Behavior* (Vol. 7, pp. 171–222). Greenwich, CT: JAI Press.

US Congress. (2003). 21st Century Nanotechnology Research and Development Act of 2003, Public Law No. 108–153.

Van de Ven, A. H. (2007). *Engaged Scholarship: A Guide for Organizational and Social Research*. Oxford, UK: Oxford University Press.

Van de Ven, A. H., & Poole, M. S. (1995). Explaining development and change in organizations. *Academy of Management Review, 20*(3), 510–540.

Von Schomberg, R. (2011). Prospects for technology assessment in a framework of responsible research and innovation. In M. Dusseldorp and R. Beecroft (Eds.), *Technikfolgen abschätzen lehren: Bildungspotenziale transdisziplinärer Methoden*. Wiesbaden: VS Verlag.

Wernerfelt, B. (1984). A resource-based view of the firm. *Strategic Management Journal, 5*(2), 171–180.

Zollo, M., & Winter, S. G. (2002). Deliberate learning and the evolution of dynamic capabilities. *Organization Science, 13*(3), 339–351.

# Chapter 7

# Socialis Commodis and Life Cycle Analysis: A Critical Examination of Uncertainty

**David M. Berube**
*Department of Communication, Public Communication of Science and Technology (PCOST) Project, North Carolina State University , North Carolina, USA*
dmberube@ncsu.edu

The Italian philosopher Giambattista Vico is best known for comparing methods on truth finding in philosophy and science. In an earlier work referenced here, he discussed briefly "socialis commodis" as the social life we live full of advantages we enjoy [1]. He argues humanity may simply have no relationship to the truth as it is transcendent. Nonetheless, even the selfish human establishes an intelligent order of sorts. Striving for intelligent order we devise methods like life cycle analysis but plagued by things like uncertainty as imperfect beings we sometimes fail miserably.

Life cycle analysis (LCA) claims to be a holistic and comprehensive way of examining and evaluating whether phenomena possess a problematic environmental health and safety footprint. Conceptually, it is hardly refutable on its face. However, proponents of LCA have

*Emerging Technologies: Socio-Behavioral Life Cycle Approaches*
Edited by Nora Savage, Michael Gorman, and Anita Street

ISBN 978-981-4411-00-4 (Hardcover), 978-981-4411-01-1 (eBook)
www.panstanford.com

done a less than admirable job finding ways to incorporate the common interests of society into their algorithms. This chapter will examine three classes of challenges focusing on nanotechnology and emerging technologies.

First, how is it possible to develop a diagnostic like LCA when the levels of uncertainty associated with emerging technologies, in this instance nanotechnology, remain so high? Populating an LCA with meaningful data is challenged by what remains unknown about nanotechnology. This problem is not limited to what we know and do not know about the toxicology of nanoparticles, although this receives a lot of attention. The intrinsic interdisciplinary nature of the fields composing nanoscience and nanotechnology may perpetuate this problem as well. Uncertainty is especially problematic when we attempt to include data about societal interests some of which is qualitative and some quantitative.

Second, deciding where to include data about societal interests in an LCA has not been established in the prevailing LCA models. While societal concerns can be justified at different points in LCA, we must decide whether societal interests belong at every major decision node in an LCA or at some particular points and not others. While it is politically correct to argue societal interests are important at every point, locating debates over social dimensions at every node might be self-defeating. Saying social dimensions data belong everywhere in an LCA model may result in it included nowhere. And, if experience serves as a guide, as LCA is designed it is imperative we determine how and when societal interests function in the analysis, else societal interests become an afterthought. Similar to the relegation of social scientists to the concluding session in conferences on nanotechnology, failing to plan for the inclusion of societal interests data into LCA may relegate it to window-dressing, which might complicate the relationship between nanotechnology scientists and the public. We know public reluctance about engagement activities, like consensus conferences, has been dominated by relevance issues. If the public perceives its participation is predominantly symbolic, the gap between science and engineering and public understanding will widen further.

Third, LCA like most other forms of risk assessment is not optimal when data is incomplete. However, risk managers have been reluctant to consider new ways of assessing risk that is not dominated by explicit and finite projections of exposure and dosage.

When risk assessment is challenged by emerging science and technology, attempting to force the phenomena under study into an algorithm that is not designed to accommodate it is a fool's game. Contemporary research in fuzzy-set theory and heuristics offers opportunities to redesign the risk assessment models to produce a stronger algorithm to examine phenomena like nanotechnology steeped in unknowns. This approach may produce a more nuanced LCA model designed to assess emerging technologies.

## Models

There are many models for life cycle analysis, but for purposes of this discussion, only one will be examined: the "Nano-Risk Framework" (NRF) [2] designed by the Environmental Defense Fund and DuPont Nano Partnership. It was not selected for its superiority over other models for most are similar. Since the authors of this framework and the author of this chapter were colleagues in the International Council on Nanotechnology, the latter is very familiar with this framework hence uses the NRF in this examination. As such, the "Nano-Risk Framework" will merely be used to track entry points for societal inputs as well as an illustration to understand the impacts of high levels of uncertainty in computing a LCA.

At first glance, LCA models have no clearly identified societal dimensions data-entry points unless societal issues are defined exclusively as human and eco-toxicological ones. As such, the premise of this chapter is that life cycle analysis has mostly ignored societal dimensions as a variable in an analysis of technology, specifically, nanotechnology.

## Analysis with Uncertainty

The NRF model demands data in six broad categories: (1) description, (2) profile, (3) net assessment, (4) management assessment, (5) decide, document, and prescribed action, and (6) review and adaptation, while simultaneously assessing, prioritizing, and generating data. NRF calls for finite data sets throughout the model, especially in the evaluation, management assessment, and decision and documentation categories. Questions of high persistence and bioavailability as well as stability and transformability come from base-set data collection. The data sets must be sufficiently finite to produce rationales for triggering calls for additional data generation or deciding enough data exists to move on. In addition, the entire health-hazard data and environmental fate data sets demand fixed data. However, almost by definition, emerging technology involves extreme unknowns. Furthermore, projections and exaggerated claims from experts and nonexperts alike will aggravate efforts to produce data points for the NRF model.

Uncertainty is the same for an expert toxicologist working out of his lab as it is for a parent trying to decide whether his daughter should sleepover at a girlfriend's house. It is mostly but not always predicated on less than optimal data. Uncertainties about environmental contaminants are often based on a sense of "not knowing" about these contaminants because they are invisible. Health problems related to exposure to contaminants may not show up for years. Information about contaminants may not have been provided by authorities [3–6].

However, uncertainty is constituted by more than merely insufficient information. On one level, efforts to reduce uncertainty can amplify uncertainty by calling for more and more directions for data generation and analysis. In addition, uncertain findings in one study can be spun to justify more research. Furthermore, research findings can have substantive impacts and at the same time open more opportunities to generate more data in turn perpetuating grant worthiness for teams of researchers as well as its associated indirect funds to their university. Uncertainty reduction has an important ideological component as well leading experts to argue data analysis can be used to manipulate uncertainty in one or more desired directions [7]. A classic illustration of uncertainty being used as an ideological tool has been efforts by some elected officials to discredit claims on climate change by highlighting findings from outside the consensus in an effort to thwart policy [8]. Research and findings associated with how risk and uncertainty intersect in the subject field of nanotoxicology may have significant implications galvanizing unease, supplying reassurance, and even resulting in delayed regulatory actions.

While nanoscience and nanotechnology may appear exotic, they are not. A strong case can be made that applied nanoscience is little more than the next step in material science. Nonetheless, nanotechnology is a textbook example of an emergent technology:

boldly inter-disciplinary, conceptually fuzzy [9], and increasingly unsettling for many experts and inexperts alike. We may conclude nanotechnology is inherently or intrinsically tentative and certainty as a finite concept may be wholly illusive.

The paradox of science is that our world has become more predictable and controllable and more complex and chaotic at the same time. Over the last decade, the rise of various new technologies that assume and actually create uncertainty (nanotechnology, genetic modification, synthetic biology) requires researchers in the field of risk studies to better examine new complexities and assumptions of risk and certainty in a way that will allow us to sufficiently plan for the decades ahead.

Uncertainty is a function of many unknowns, but as a socially constructed reality, it is affected by nonrational variables as well. While risk is clearly a function of some hazard, moderated by dosage and exposure variables, these variables themselves are also affected by the uncertainties associated with hazard, dosage, and exposure—and many of these uncertainties come directly from science. Expecting science to sufficiently reduce uncertainty such that models using finite values can predict effects may be unrealistic.

Simply put, as scientists attempt to increase certainty, contingencies are added to their claims as a hedge that has the unfortunate side effect of degrading the importance of the claims. "Because science involves producing knowledge about what was previously unknown, uncertainty is a normal and necessary characteristic of scientific work... [S]cientific uncertainty is not simply something that scientists try to eliminate through their research. Scientists also actively construct and effectively use it in scientific articles and in public science contexts" [10].

Zehr has gone as far as to argue: "Scientific work does not just reduce uncertainty, it actively constructs it... Science is as much an uncertainty generator as it is a certainty producer" [10]. To further complicate the issues, even when science adds contingencies to findings to produce a more finite set of data, as the modified conclusion "moves into more formal and more universal settings, it often loses some of its contingencies" [10].

What tools are available to address uncertainty? Traditional communication approaches toward uncertainty lie along three theoretical lines of thinking: uncertainty reduction theory [11], problematic integration theory [12,13], and uncertainty management theory [14]. All three theories postulate uncertainty as an autonomous cognitive state affected by language. In addition, several common themes emerge in this literature about what uncertainty might entail, including (1) the perception of "not knowing," (2) emotions, and (3) communication processes.

In addition, uncertainty management theory accommodates the willful obfuscation and the seeking of vague, imprecise information. Simply put, sometimes it is deemed profitable to claim uncertainty to mask a decision already made and to avoid dealing with a negative reaction. False claims of certainty exhibit confidence but may have the effect of persuading oneself of being certain. Powell *et al.* [15] pointed to several factors shaping uncertainty perceptions, which have not yet been clearly addressed, including "knowing" or perceived knowledge, the role of trust, risk judgments and issue relevance, information exposure and interpersonal discussion, and socio-demographic and contextual factors.

The fundamental assumption that more information reduces uncertainty ignores the obvious: individuals select between sources of information for many reasons, only one of which is its intrinsic quality. Other variables are at work to explain why a minister or a member of one's family may trump the opinion from a scientist—even when the information is outside the expertise of the nonscientific sources. Information can be avoided (leaving an envelope unopened), and adaptation to a state of uncertainty can even become a defense mechanism (see research on AIDS patients) [16]. More data may be insufficient to reduce uncertainty in some circumstances.

Technical communication theorists have contended uncertainty is multilayered, and as such these phenomena need to be unraveled [17]. People (experts included) are embedded in layers of context that subject them to complimentary as well as contradictory forces. Complications in the health of an environment from ambiguous symptoms, degrees of hazard, and unclear solutions can cause uncertainties in financial well-being, personal relationships, and general well-being. Finally, the spatiotemporal dimensions of uncertainties are under-investigated [18]. Uncertainties can be brief or ongoing. The level of uncertainty can increase or decrease over time as people grow accustomed to its presence. Uncertainty about one phenomenon will bleed into another. The basis for uncertainty about one element of a class of phenomena may not be universally shared by other elements in that class. In cases where there is little, if any, transference or contagion we find uncertainty anomalies, the grounds and warrants for which are important variables.

Uncertainty as a subjective state is not easily resolved by more information. There is an external and internal view of uncertainty whereby individuals attribute uncertainty to some phenomenon in the outside world while locating the source of uncertainty within the individual [19]. As Heyman *et al.* wrote, "people make probability judgments based on what they know as much as or more so than on what they don't know" [20]. As such, these observations may help explain some discrepancies in opinion studies about nanotechnology to date.

Trust may be as important as data. Powell *et al.*'s tag of trust as a variable in uncertainty is worth further testing. As Johnson has noted, "It is still unclear whether uncertainty in risk estimates is an independent issue in risk communication, or whether attitudes toward it merely reflect other attitudes, such as whether one can control risks to oneself or trusts the source of uncertain risk estimates" [21].

Slovic and others [22] have demonstrated that the power of risk judgments is related to perceived uncertainties about risks. The more important the risk, presumably, the more willing an individual may be to reduce their uncertainty about the risk. However, this phenomenon is much more complex than it seems, given high-risk populations who tend, at times, to make little, if any, effort to reduce their uncertainty regardless of the risks involved, a phenomenon that surfaces regularly in health communication and deadly diseases.

Socio-cultural experiences, including formal education, disposable income, and so forth, might be related to understanding of scientific information. In addition, there still remain too many instances where privilege is correlated to uncertainty. While much of this research should be applauded, the geography of risk and uncertainty studies remains relatively flat at this time. The construction of uncertainty has many actors working on many different levels, using a wide array of tools or strategies. In addition, research to date does not fully appreciate the clash between organizational cultures and how the tension and intersectional sensibilities of experts complicate the construction of effective risk messages when addressing issues with moderate to high levels of uncertainty.

Finally, thinking about risk and uncertainty seems limited by an uncertainty reduction paradigm. Deciding which variables should receive more attention than others and whether one or more should be discounted altogether remains highly disputable. At times, we may need to simply accept the thesis that uncertainty cannot always be reduced and that our efforts to reduce uncertainty at nearly all costs may be pathological [23]. Associatively, the conceptual construction of uncertainty rings differently to different stakeholders. How a regulator approaches uncertainty seems at odds with how the inexperts approach uncertainty. We must also consider that uncertainty can be a tool used by some stakeholders to forward their own agendas, often self-serving but sometimes altruistic but misinformed. Recall that uncertainty can be introduced as a tactic to garner additional resources. For example, toxicologists seldom, if ever, admit there exists sufficient data to make a reasonably certain decision. As we have often seen in the nanotechnology literature, some groups have vested self-interests in maintaining high levels of perceived risk and others benefit from claims of high levels of uncertainty [23]. Some stakeholders emphasize uncertainty as much as risk to amplify a hazard, raise the visibility of their organization or issue, and shake loose heightened levels of constituent funding. Scientists use uncertainty to protect themselves against charges of dogma and bias and as a technique to provide a boundary between themselves and policy makers they advise [10]. Uncertainty is a rich and dense construct and needs another examination by scholars who will approach uncertainty as something not necessarily resolved by more data.

Lipschitz and Strauss asked an important question a decade ago: "Are there systematic relationships between different conceptualizations of uncertainty and different methods of coping?" [24]. Uncertainty should not be conceptualized as an umbrella term under which all unknowns are stored. If all uncertainties are not equal, then knowing what we may be up against may enable us to fashion an appropriate response.

Therefore, uncertainty has to be more than not knowing. Most people make efforts to manage their uncertainty rather than eliminate it altogether with more and more data. Since uncertainty reflects individual-level emotions and cognitions as well as social and contextual factors, including (1) the content of perceived knowledge about risks; (2) social perceptions (e.g., trust in government, perceived risk to others); (3) perceived issue importance; (4) interactions with mediated information and interpersonal discussions; and (5) socioeconomic and contextual factors [15], eliminating uncertainty may beckon failure.

Approaches to examining uncertainty and risk from a decision-making perspective typically and popularly include Subjective-Expected-Utility [25] calculation to games involving Monte Carlo simulations [26]. However, the bulk of this literature seems to relate decision-making strategies to a version of framing (whether a situation is presented as an opportunity or a threat) [27] and to outcome history. In outcome history, a person–situation-interaction characteristic is defined as the degree to which the decision maker believes that previous risk-related decisions have resulted in successful or unsuccessful outcomes [28]. These approaches are coping strategies rather than certainty directed decision-making strategies.

Of course, decision makers must make conclusions. As such, they tend to use one or more of five strategies to cope with uncertainties standing in their way: reducing uncertainty, assumption-based reasoning, weighing pros and cons of competing alternatives, suppressing uncertainty, and forestalling [24]. We know much less about how decision-making organizations cope, especially in cases when the process remains unstructured [29]. There is some evidence for when people cannot find a way to reduce the uncertainty in their lives, they make decisions regardless of uncertainty by employing nonrational and irrational criteria (shortcuts or heuristics), change the ways they make decisions, or build a cocoon of uncertainty to protect themselves from the complexity of the world out there [30]. Similarly, when experts confront uncertainty, they tend to increase or improve the data flow; reclassify data that once was accepted to discount it or was once rejected in order to accommodate it; or reject a whole way of thinking, opting for a new paradigm.

Public communication is troubled by uncertainty as well. The context in which scientists make scientific claims before nonscientists has been touted public science and seems to jointly emerge from scientists who are interviewed and from writers themselves [10]. As such, public science has animated, if not amplified in the Kasperson [31] sense, the association of risk and uncertainty with science and technology. For example, quite regularly, in an attempt to appear balanced, reporters provide podiums for individuals advancing untenable claims; or, alternatively, certain basic assumptions may be simply overlooked. As Zehr has argued, "in the more limited space of public science, sometimes tacit knowledge may be left out, leading to an appearance of contradictory claims" [10]. Indeed, given the numerous and contradictory purposes of public science, "uncertainty is a much more common feature of public science than it might appear to be on the surface, and it could be intrinsic to public science itself" [20].

More risk studies are examining the media [32–35], and many projects of public engagement are testing the role communication can play in risk estimation. At the same time, another research has shown that inexperts, by and large, do not aggressively seek scientific information [36]. All of this suggests inexpert audiences pose special challenges. They tend to be more heterogeneous sources, for example, and they seek more ephemeral and gray sources for amplification and attenuation, such as the web or popular media. Public science typically features multiple speakers with multiple versions of knowledge, and this cacophony of sorts can result in a range of challenges.

Scientists may find the appearance of uncertainty managed by other parties, such as the media or civic advocacy groups to ends contrary to those of the scientists. Behavior of this sort tends to produce uncertainty spirals, uncertainty out of control.

## Analysis without Societal Dimensions

Once again, the NRF model demands data in six broad categories: description, profile, net assessment, management assessment, prescribed action, and review and adapt, all the while assessing, prioritizing, and generating data. Presumably as more data surfaces and as values in each of the broad categories change, (re)iteration takes place. While iteration is a powerful dynamic to resolve data discrepancies over time, there is no provision to (re)iterate as social dimensions data improves. Except for the category "use and reuse" under profile, there are no other input points where direct interaction with consumers and where other social dimensions data can contribute to LCA.

The public's role in science and technology is questionable. Engagement theorists claim the public has a right to participate in decision making affecting the common good. While politically correct, it does not address a few issues. Some publics are not interested, regardless of how much they are informed, and they remain uninterested. Some publics have deferred their input to their elected officials and the bureaucrats the officials hire. Some publics exaggerate their participation by claiming to speak for others and so forth. On another level, engagement theorists claim the public needs to participate as consumers, otherwise technology does not become profitable and is shelved. Of course, some publics are difficult to please and do not make objective purchasing decisions. For example, we consume much of the wrong stuff, especially food, and in wrongful amounts growing fatter and dying younger.

Another consideration is drawn from engagement theorists who argue the public will express their unhappiness by boycotting technologies calling upon their elected officials to publicly defund suspected technologies. Too often, the public sees science and technology locked in a could/should conundrum. Publics perceive scientists and technologists deciding to do research because science has given them the tools to research effectively. Normative questions about what should be done are sometimes unasked. This conundrum is foisted onto scientists and technologists by the rewards system that populates their professions. First out of the gate is the first rewarded. Finally, elected officials have convinced themselves the public pays attention to science and technology policy creating a perception of public perception driver for societal-dimensions research as well as a call for public participatory governance. Mostly fed by proclamations from self-appointed standard bearers for the public (non-governmental organizations), one is hard pressed to cite a single time when anyone was voted out or into office on a science and technology issue.

Societal dimensions, often called societal implications, have their own budget line in the federal funding requests under the National Nanotechnology Initiative (NNI). The National Human Genome Research Institute's (NHGRI) Ethical, Legal and Social Implications (ELSI) Research Program was established in 1990 as an integral part of the Human Genome Project (HGP) to foster basic and applied research on the ethical, legal and social implications of genetic and genomic research for individuals, families, and communities [37]. When the NNI was proposed, it was deemed similar research may be necessary in nanotechnology, given the hyperbolic rhetoric associated with exotic nanotechnological applications as well as the repeated claim nanotechnology was the "next industrial revolution" [38].

Societal dimensions of science and technology include a plethora of concerns. Back in 2001, societal dimensions/implications were defined as "educational, technological, economic, medical, environmental, ethical, legal, cultural, etc." [39]. The etcetera is telling enough. Societal implications are an open set, and given a set of good definitions and well-placed arguments almost anything might find itself under the societal dimensions umbrella.

Societal dimensions of nanotechnology include (1) economic effects, such as employment consequences and shifting demands on resource markets, here and abroad; (2) privacy associated with more intrusive computer-supported surveillance and data management; (3) justice involving outsourcing hazardous activities to those least able to protect themselves while generating applications for those with the wherewithal to purchase expensive therapies and products; (4) defense related issues involving nano-soldiering and fighting wars with lower human costs for one or both sides all of which might make war-fighting less unlikely as a last resort; and (5) dehumanization associated with more ubiquitous technologies of all sorts, including those associated with human enhancement. Nanotechnology with its hyperbolic claims and its revolutionary potentials seems to aggravate all of these.

## Economy

Globalization has challenged the public understanding of secured employment both in the United States and developing economies. Nanotechnology promises to reduce demands on some natural resources impacting markets that draw from the physical world around them to produce value. On the other hand, nanotechnology, at least in the short run, may draw heavily from rare earth minerals inflating value there. Employment is a powerful driver in the lives of almost everyone. Economic claims about nanotechnology do not seem to impact individuals who scratch together a meager existence. Benefits to developing economies, if realized, will come after a substantial lag time. This lag will see disruptions in many different markets and across many economies [40,41].

## Privacy

Quicker and more powerful computational resources fostered by nanotechnological discoveries will increase the capacity to manage huge sums of data about individuals, organizations, industry, etc. Efforts to secure databanks have been mixed. Talks of cyberwars may grab headlines, but for most of the public, the concern centers on identity theft, personal records (especially medical records), and so forth [42].

## Justice

The primary beneficiaries of nanoproducts will be those who can afford to purchase them. If there are risks associated with production of nanoparticles, we can expect facilities to relocate where the labor force is less expensive and where regulations regarding production are weaker. History seems to find rich–poor gaps part and parcel of technological revolutions [43].

## Defense

One of the largest categories of government-funded research in nanotechnology is defense. Nano-soldiering involves lightening loads carried into combat, improving battle suits to protect wearers, increasing resilience while reducing the weight of war fighting vehicles, developing new and more sophisticated command-control computational capabilities, and so forth. Some even suggest a new generation of weapon systems that remove the human-damage factor from the decision whether to invade or not [44]. Sea power and air power began this race. Advanced robotics and computation may take it to new levels. Many are concerned that without the human factor as a variable for all the sides in a conflict, the war itself will become more likely [45].

## Dehumanization

The public has been exposed to claims from nanomedicine, nanoceuticals, and others that nanotechnology offers opportunities to extend human lifespans and improve health and well-being. Lifespans may lengthen as people regularly receive nanotechnology-facilitated implants. More hyperbolically, sportsmen could excel with elective implants. Transhumanists may merge with the machine. Posthumanists will move deeper into cyberhood. While some extropians might find themselves entertained by the prospects, for most of the public what emerges may be dehumanizing. Dehumanization has been a plague on humanity. When people become things, they become dispensable. When people are dispensable, any and every atrocity can be justified. Once justified, they seem to be inevitable for every epoch has evil, and dehumanization is evil's most powerful weapon. Humanity is the essential defining feature of existence [46,47].

For the public, issues like those summarized above remain important and deserve to be aired in debates over nanotechnology. Unfortunately, LCA models reflect the prioritization given to these values by those empowered to make decisions. In too many instances, societal dimensions are treated as less important to technical dimensions, especially in terms of budgeting.

For example, one societal dimension is environmental consequences and environmental health and safety, a phrase that has grown to include both human and ecological toxicity. Since it is budgeted separately from other societal dimensions, it is not the subject of the following paragraph. Nonetheless, it is important to note many experts mostly tend to categorize the societal dimension of nanotechnology as a toxicological phenomenon.

In terms of budgets, societal dimensions and education were lumped together for years. The requested 2012 budget for education and societal dimensions was $69 million [48], which was reduced quickly to $29 million. Once education was removed from the societal dimensions (ELSI) category, the FY 2012 request for societal-dimensions research became $5.28 million out of an overall request for $455.95 million [49], accounting for 1.1 percent of the budget. With only 1.1% of the NNI research budget allocated to societal dimensions data generation and analysis, the LCA models may simply reflect the importance associated with societal dimensions.

## Analysis with Methodological Creativity

LCA like most other forms of risk assessment is not optimal when data is incomplete. However, risk managers and risk-assessment theorists have been reluctant to consider new ways of assessing risk, which are not dominated by explicit projections of exposure, dosage, and hazard. When risk assessment is challenged by emerging science and technology, attempting to force the phenomena under study into an algorithm that is not designed to accommodate will be counterproductive. Contemporary research in fuzzy-set theory and heuristics offers opportunities to rewrite the risk-assessment models to produce stronger algorithms to examine phenomena like nanotechnology riddled with unknowns.

Another source of uncertainty has much less to do with insufficient data and more to do with a deficient methodology for assessing the available data. LCA, like almost all other forms of risk assessment, demands finite data values. Societal-dimensions data does not tend to be finite but rather describes a range of phenomena, which while reducible to quantifiable values are seldom reducible to a single data point.

Stochastic techniques defined as the "scientific reasoning model" are based on probability theory and the fuzzy-set techniques known as a "human reasoning model" are based mostly on linguistic parameters and qualitative values. Used complementarily, they offer opportunities to reduce uncertainty as well as manage the transition to a more relevant risk assessment/management life cycle algorithm for emerging science and technology, including nanotechnology.

Before there was fuzzy-set theory, statisticians employed the "scientific-reasoning model" to consider incomplete data in random uncertain viewpoints. "The empirical Bayes method [50] as a practical statistics tool provided a way in which similar historical data could be used to improve the current results which come from a few experiments. Unfortunately, a fundamental problem in Bayes approach and related techniques is that in many cases there is no historical data or a prior distribution for use" [51].

Advances in the "scientific-reasoning model" involved a series of techniques to improve on probability-theory-driven solutions. The primary technique has been the use of analytical methods. This approach works best when an explicit formula is available and input parameters are clear and independent. The response-surface-replacement approach used in computer modeling involves substituting a response-surface replacement for all uncertainty and sensitivity analyses. Differential-sensitivity analysis is used when there is an agreed upon relationship between output and input variables. Still another is the well-known Monte Carlo simulation, the iterative process whereby single values are generated and used in subsequent model calculations producing a body of model calculations from which statistical results are generated. Latin hypercube sampling has been used to reduce the model calculations of the Monte Carlo method. All of these approaches have strengths and weaknesses and are hardly a full set of options in certainty reduction in risk analysis.

Enter fuzzy-set theory [52]. A fuzzy set is a class of objects with a continuum of grades of memberships. Traditionally, such a set is characterized by a membership (characteristic) function that assigns to each object a grade of membership ranging between zero and one.

Smithson and Verkuilen [53] define fuzzy-set theory as "a mathematical framework for handling categories that permit partial membership (or membership in degree) . . ." They claim "adding fuzzy sets to the social science toolbox . . ." allows us to "handle vagueness systematically" (p. 3). Huang [51] claimed fuzzy-set theory would allow us to better understand how to make sense of the uncertainties produced by the vagaries of natural language. And risk analysis and management is about communicating our findings to different publics, and expert and inexpert are all bound by one or more linguistics traditions.

Cho, Choi, and Kim [54] argued the fuzzy-number concept [55–57] can generalize both concepts of real numbers and close intervals to handle ambiguity in modeling subjective-probability judgment. They developed fuzzy-membership curves using the fuzzy-number concept to consider uncertainties of risk events on the basis of existing studies and demonstrated an example of the proposed fuzzy-membership curve of "close to."

Huang and Moraga [58] use a fuzzy controller and its set of fuzzy if–then rules to approximate functions. They claim fuzzy control is increasingly seen as a universal approximator [59] by the control community and thus is strongly used for approximating functions [60].

Consider fuzzy-set theory was employed as a component in a hybrid approach toward estimating the risks associated with an ammonia storage facility [61]. In addition, fuzzy-set theory in conjunction with the Monte Carlo method was applied to human exposure via vegetable consumption to cadmium present in surficial soils of an industrial site [62]. Darbra, Eljarrat, and Barceló [63] reviewed stochastic and fuzzy-set techniques commonly used to accommodate uncertainties associated with risk-modeling inputs and outputs. They suggest both methodologies can be engaged to deal with uncertainty and provide applications of these approaches to environmental risk assessment. They conclude "A combined approach may therefore be the best solution to deal with the uncertainties. Treating each parameter according its nature and information available could help to obtain a more realistic result for risk estimation" (p. 384).

Fuzzy-settheoryhasbeenusedbyscholarsinattemptingtounravel highly uncertain phenomena. It has been used to better understand the vagaries of natural language [51] to handle tricky dose–response relationships [64], to enlighten health assessment [65, 66] as well as environmental decision making [67]. In addition, fuzzy-set theory has been used in many different risk assessment settings involving high levels of uncertainty: dredging [68], flood control [69], nitrate contamination [70], highway-bridge safety [71], landfill pollution [72], underground storage tank pollution [73], pesticide pollution [74], soil and sediments pollution [75], earthquake prediction [76], agriculture and flooding [77], groundwater contamination [78], river water quality management [79], hydrocarbon contamination [80, 81], and potable water [82].

According to Quelch and Cameron [83], "the novel approach of fuzzy sets is an appropriate technique due to its non-statistical nature and that the amount of computation required is substantially reduced compared to the traditional probabilistic approach. The versatility of fuzzy set theory suggests that this approach could also be used to quantify other types of uncertainty present in the risk assessment process, including model uncertainty and expert opinion" (p. 463).

The use of fuzzy sets as inputs to these models is often more representative of our true knowledge of the parameters than a single value or a probability-density function [83].

LCA, as well as, risk assessment in general, may benefit from re-examining how risk assessment has been done for decades. The model was derived from more naive times. While fuzzy-set theory is not *sine qua non* for the uncertainties about emerging technology and the dearth of social dimensions input into current models, it moves us toward expression of risk more in tune with the state of scientific and social-scientific research objectives. By incorporating tools that enable a set of ranges of viable data sets rather than finite or incomplete data sets, we may complete a more viable assessment of phenomena like emerging technologies.

"The most important output from risk assessment is the capacity to provide the basis of a decision-making process. The results of such decisions should be presented to the environmental managers and public in plain language and in line with the way humans think, rather than as difficult numbers or calculations... In any case, there is no doubt that, for concepts, such as "safe", "tolerable" and "severe", boundaries are very hard to define. They are linguistic statements, and it is not easy to assign numerical values to them. But curiously, we make this kind of judgment every day in our lives without realizing it—"Driving fast is very dangerous" – "this behavior is intolerable"—"this is safe" [63].

## Conclusion

A life cycle–analysis technique for emerging technologies, especially nanotechnology, is preferable to the less comprehensive models tending to undervalue end-of-life considerations. However, much too often, contemporary risk assessment calls for best guesses when data is incomplete. Best guesses belong in carnival side shows and not regulatory policy making. Bad guesses hold the scientific and regulatory community up for ridicule. Furthermore, getting it wrong can be haunting and may set back regulatory initiatives for decades. In the interim, a technological development may run roughshod over health and overall well-being of the public. LCA must address two foundational problems: how to resolve the uncertainty issues and how to include societal dimensions in their algorithms. Finally, given the extreme set of unknowns associated with emerging technologies, it might be in our best interests to examine the risk-analysis methodologies upon which we have grown dependent. Fuzzy-set theory offers new insights in risk analysis and when coupled with LCA may provide a comprehensive and versatile model to regulate nanotechnology, synthetic biology, and whatever is next in line. To harken back to Vico, we remain imperfect beings making imperfect arguments. Whether we get closer to truth is unclear and whether LCA as method offers an increased likelihood of truth finding demands immense humility. The uncertainty endemic to emerging sciences requires us to unlearn our native language of scientific certainty and return to the pristine beggary of words Vico advocated to have any chance of arriving at truth and even then Vico would be suspicious of the claims of the devotee wordsmiths, in this case algorithm builders.

## Acknowledgment

This work has been supported in part by the National Science Foundation (SBE 0809470). The opinions expressed are the author's and may not be shared in whole or in part by the NSF and North Carolina State University.

## References

1. Vico, G. (1709, 1840). De Constantia Philologiae [On the Constancy of Philology]. In J. Michelet, Ouvres Choisies de Vico (2 vols.). Brussels: Meline, Can et Compagnie.
2. Nanorisk Framework. (2007). http://www.nanoriskframework.com/page.cfm?tagID=1095 (Accessed January 29, 2008).
3. Adeola, F. O. (2000). Endangered community, enduring people: toxic contamination, health, and adaptive responses in a local context, *Environment and Behavior*, 32(2), 209–249.
4. Brown, P., & Mikkelson, E. (1990). *No Safe Place: Toxic Waste, Leukemia, and Community Action*. Berkeley, CA: University of California Press.
5. Hallman, W. K., & Wandersman, A. (1992). Attribution of responsibility and individual and collective coping with environmental threats, *Journal of Social Issues*, 49(4), 101–118.
6. Irwin, A., & Wynne, B. (1996). *Misunderstanding Science? The Public Reconstruction of Science and Technology*. Cambridge: Cambridge University Press.
7. Brashers, D. E., Neidig, J. L., Haas, S. M., Dobbs, L. K., Cardillo, L. W., & Russell, J. A. (2000). Communication in the management of uncertainty: The case of persons living with HIV or AIDS, *Communication Monographs*, 67, 63–84.
8. McCright, A. M., & Dunlap, R. E. (2003). Defeating Kyoto: The conservative movement's impact on U.S. climate change policy, *Social Problems*, 50(3), 348–373.
9. Chen, S. J., & Hwang, C. L. (1992). *Fuzzy Multiple Attribute Decision Making: Methods and Application*. Berlin: Verlag Springer.
10. Zehr, S. (1999). Scientists' representation of uncertainty. In M. Friedman, S. Dunwoody, and C. Rogers (Eds.), *Communicating Uncertainty: Media Coverage of New and Controversial Sciences* (pp. 3–21). Mahwah, NJ: Lawrence Erlbaum Assoc.

11. Berger, C. R., & Calabrese, R. J. (1975). Some explorations in initial interaction and beyond: toward a developments theory of interpersonal communication, *Human Communication Research*, 1, 99–112.
12. Babrow, A. S. (1995). Communication and problematic integration: Milan Kundera's 'lost letters' in the book of laughter and forgetting, *Communication Monograph*, 62, 284–300.
13. Babrow, A. S. (2002). Communication and problematic integration: understanding diverging probability and value, ambiguity, ambivalence, and impossibility, *Communication Theory*, 2, 95–130.
14. Brashers, D. E., Neidig, J. L., Haas, S. M., Dobbs, L. K., Cardillo, L. W., & Russell, J. A. (2000). Communication in the management of uncertainty: the case of persons living with HIV or AIDS, *Communication Monographs*, 67(1), 63–84.
15. Powell, M., Dunwoody, S., Griffin, R., & Neuwirth, K. (2007). Exploring lay uncertainty about an environmental health risk, *Public Understanding of Science*, 16, 323–343. Doi: 10.1177/0963662507074491.
16. Mishel, M. H. (1990). Reconceptualization of uncertainty in illness theory, *Image: Journal of Nursing Scholarship*, 22, 256–262.
17. Brashers, D. E., & Babrow, A. S. (1996). Theorizing communication and health, *Communication Studies*, 47, 243–251.
18. Babrow, A. S., & Kline, K. N. (2000). From 'reducing' or 'managing' uncertainty: reconceptualizing the central challenge in breast self-exams, *Social Science and Medicine*, 51, 1805–1816.
19. Kahneman, D., & Tversky, A. (1982). Variants of uncertainty, *Cognition*, 11, 143–157.
20. Heyman, B., Henrikesen, M., & Maughan, K. (1998). Probabilities and health risks: A qualitative approach, *Social Science and Medicine*, 47, 1295–1306.
21. Johnson, B. B. (2003). Further notes on public response to uncertainty in risks and science, *Risk Analysis*, 23(4), 781–789.
22. Johnson, B. B., & Slovic, P. (2006). Presenting uncertainty in health risk assessment: Initial studies of its effects on risk perception and trust, *Risk Analysis*, 15(4), 485–494.
23. Berube, D. (2008). Intuitive toxicology: the public perception of nanoscience. In F. Allhoff, & P. Lin (Eds.), *Nanoethics: Emerging Debates* (pp. 91–108). London: Springer.
24. Lipscihtz, R., & Strauss, O. (1997). Coping with uncertainty: A naturalistic decision-making analysis, *Organizational Behavior and Human Decision Processes*, 69(2), 149–163.

25. Wakker, P. (1993). Savage's axioms usually imply violation of strict stochastic dominance, *The Review of Economic Studies*, 60(2), 487–494.

26. Davis, T. J., & Keller, P. (1997). Modeling uncertainty in natural resource analysis using fuzzy sets and Monte Carlo simulation slope stability prediction, *International Journal of Geographical Information Science*, 11(5), 409–434. Doi: 10.1080/136588197242239.

27. Jackson, S. E., & Dutton, J. E. (1988). Discerning threats and opportunities, *Administrative Science Quarterly*, 33, 370–387.

28. Sitkin, S., & Weingart, L. (1995). Determinants of risky decision-making behavior: A test of the mediating role of risk perceptions and propensity, *The Academy of Management Journal*, 38(6), 1573–1592.

29. Mitzburg, H., Raisinghani, D., & Théorêt, A. (1976). The structure of unstructured decision processes, *Administrative Science Quarterly*, 21(2), 246–275.

30. Merry, U. (1985). *Coping with uncertainty: Insights from the new sciences of chaos, self-organization, and complexity*. Westport, CT: Praeger.

31. Kasperson, R. E., Ortwin, R., Slovic, P., Brown, H., Emel, J., Goble, R. L., *et al.* (1988). The social amplification of risk: A conceptual framework, *Risk Analysis*, 8, 177–187.

32. Coleman, C. (1993). The influence of mass media and interpersonal communication on societal and personal risk judgments, *Communication Research*, 20(4), 611–628.

33. Friedman, S. M., Dunwoody, S., & Rogers, C. L. (Eds.) (1999). *Communicating Uncertainty: Media Coverage of New and Controversial Science*. Mahwah, NJ: Lawrence Erlbaum Associates.

34. Hansen, A. (1993). *The Mass Media and Environmental Issues*. New York: Leicester University Press.

35. Hornig Priest, S. (2001). *A Grain of Truth: The Media, the Public, and Biotechnology*. New York: Rowman and Littlefield.

36. Bush, J., Moffatt, S., & Dunn, C. (2001). Keeping the public informed? Public negotiation of air quality information, *Public Understanding of Science*, 10, 213–229.

37. The Ethical, Legal and Social Implications (ELSI) Research Program. (2011). http://www.genome.gov/ELSI/ (Accessed 21 June, 2011).

38. White House Press Release. (2000). *National Nanotechnology Initiative: Leading to the Next Industrial Revolution*, http://clinton4.nara.gov/WH/New/html/20000121_4.html (Accessed 21 June, 2011).

39. Roco, M. C. (2001). Introduction. In W. S. Bainbridge (Ed.), *Societal Implications of Nanoscience and Nanotechnology* (pp. 1–3). Boston: Springer/Kluwer Academic Publishers.

40. Feller, I. (2001). An economist's approach to analyzing the societal impacts of nanoscience and nanotechnology *Societal Implications of Nanoscience and Nanotechnology: NSET Working Report* (pp. 108–114). Arlington, VA: WTEC.

41. Armstrong, T. O. (2008). Nanomics: The economics of nanotechnology and the Pennsylvania initiative for nanotechnology, *Pennsylvania Economic Review*, 16(1), 1–19.

42. Lewenstein, B. V. (2006). What counts as a "social and ethical issue" in nanotechnology? In J. Schummer & D. Baird (Eds.), *Nanotechnology Challenges: Implications for Philosophy, Ethics and Society* (pp. 201–216). Singapore: World Scientific Publishing Company.

43. Invernizzi, N., & Foladori, G. (2005). Nanotechnology and the developing world: Will nanotechnology overcome poverty or widen disparities? *Nanotechnology Law and Business Journal*, 2(3), 1–10.

44. Leahy, M. (2010). Keeping up with the drones: Is just war theory obsolete? *Online Information for the Defense Community*. http://www.dtic.mil/dtic/tr/fulltext/u2/a526187.pdf. (Accessed October 22, 2012).

45. Ratner, D., & Ratner, M. A. (2004). *Nanotechnology and Homeland Security: New Weapons for New Wars*. Upper Saddle River, NJ: Prentice Hall.

46. Nordmann, A. (2004). Social imagination for nanotechnology, in *Nanotechnologies: A Preliminary Risk Analysis on the Basis of a Workshop Organized in Brussels on 1–2 March 2004 by the Health and Consumer Protection Directorate General of the European Commission*, European Community, Brussels, pp. 111–114.

47. Berube, D. M. (2006). *Nanohype: The Truth Behind the Nanotechnology Buzz*. Amherst, NY: Prometheus Books.

48. Roco, M. C. (2011). *National Nanotechnology Investment in the FY 2012 Budget*. www.aaas.org/spp/rd/rdreport2012/12pch23.pdf (Accessed June 21, 2011).

49. National Nanotechnology Initiative/NSF-Wide Investments. (2011). *FY 2012 NSF Budget Request to Congress*, www.nsf.gov/about/budget/fy2012/pdf/35_fy2012.pdf (Accessed June 21, 2011).

50. Maritz, J. S., & Lwin, T. (1989). *Empirical Bayes Methods*. New York: Chapman & Hall.

51. Huang, C. F. (2007). Risk analysis with information described in natural language, In Y. Shi, *et al.* (Eds.), *Computational Science ICCS 2007, Part III, Lecture Notes in Computer Science* 4489 (pp. 1016–1023). Doi: 10.1007/978-3-540-72588-6_164.

52. Zedah, L. (1965). Fuzzy sets, *Information and Control*, 8, 338–353.

53. Smithson, M., & Verkuilen, J. (2006). *Fuzzy Set Theory: Application in the Social Sciences*. Thousand Oaks, CA: Sage Publications.

54. Cho, H. N., Choi, H. H., & Kim, Y. B. (2002). A risk assessment methodology for incorporating uncertainties using fuzzy concepts, *Reliability Engineering and System Safety*, 87, 173–183.

55. Dubois, D., & Prade, H. (1982). The use of fuzzy numbers in decision analysis. In M. M. Gupta & E. Sanchez (Eds.), *Fuzzy Information and Decision Processes* (pp. 309–321). New York: North-Holland.

56. Dubois, D., & Prade, H. (1989). Decision evaluation under uncertainty and imprecision. In J. Kacprzyk, & M. Fedrizzi (Eds.), *Combining Fuzzy Imprecision with Probabilistic Uncertainty in Decision Making* (pp. 48–64). Berlin: Springer, Berlin.

57. Watson, T. S. R., Weiss, J. J., & Donnell, M. L. (1979). Fuzzy decision analysis, *IEEE Transactions on Systems, Man, and Cybernetics*, 9(1), 1–9.

58. Huang, C. F., & Moraga, C. (2005). Extracting fuzzy if–then rules by using the information matrix technique, *Journal of Computer and System Sciences*, 70, 26–52.

59. Verbruggen, H. B., & Bruijn, P. M. (1997). Fuzzy control and conventional control: what is (and can be) the real contribution of fuzzy systems? *Fuzzy Sets and Systems*, 90(2), 151–160.

60. Dubois, D., & Prade, H. (1997). The three semantics of fuzzy sets, *Fuzzy Sets and Systems*, 90(2), 141–150.

61. Quelch, J., & Cameron, I. T. (1994). Uncertainty representation and propagation in quantified risk assessment using fuzzy sets, *Journal of Loss Prevention in the Process Industries*, 7(6), 463–483.

62. Guyonnet, D., Bourgine, B., Dubois, D., Fargier, H., Côme, B., & Chilès, J. P. (2003). Hybrid approach for addressing uncertainty in risk assessments, *Journal of Environmental Engineering*, 129(1), 68–78. Doi: 10.1061/(ASCE)0733-9372(2003)129:1(68).

63. Darbra, R. M., Eljarrat, E., & Barceló, D. (2008). How to measure uncertainties in environmental risk assessment, *Trends in Analytic Chemistry*, 27(4), 377–385. Doi: 10.1016/j.trac.2008.02.005.

64. Bárdozzi, A., Bogárdi, I., & Duckstein, L. (1993). Fuzzy nonlinear regression analysis of dose-response relationships, *European Journal of Operational Research*, 66, 36–51.

65. Kentel, E., & Aral, M. M. (2005). 2D Monte Carlo versus 2D Fuzzy Monte Carlo health risk assessment. *Stochastic Environmental Research and Risk Assessment*, 19, 86–96. Doi: 10.1007/s00477-004-0209-1.

66. Kentel, E., & Aral, M. M. (2007). Risk tolerance measure for decision-making in fuzzy analysis: A health risk assessment perspective. *Stochastic Environmental Research and Risk Assessment*, 21, 405–417. Doi: 10.1007/s00477-006-0073-2.

67. Tesfamariam, S., & Sadiq, R. (2006). Risk-based environmental decision-making using fuzzy analytic hierarchy process (F-AHP), *Stochastic Environmental Research and Risk Assessment*, 21, 35–50. Doi: 10.1007/s00477-006-0042-9.

68. Lee, Y. W., Bogardi, I., & Stansbury, J. (1991). Fuzzy decision making in dredged-material management, *Journal of Environmental Engineering*, 117(5), 514–630.

69. Esogbue, A. O., Theologidu, M., & Guo, K. (1992). On the application of fuzzy sets theory to the optimal flood control problem arising in water resources systems, *Fuzzy Sets and Systems*, 48, 155–172.

70. Lee, Y. W., Dahab, M. F., & Bogardi, I. (1995). Nitrate-risk assessment using fuzzy-set approach, *Journal of Environmental Engineering*, 121(3), 245–256. Doi: 10.1007/s11270-005-9005-x.

71. Sobanjo, J. O., Stukhart, G., & James, R. W. (1994). Evaluation of projects for rehabilitation of highway bridges, *Journal of Structural Engineering*, 120(1), 81–99.

72. Hagemeister, M. E., Jones, D. D., & Woldt, W. E. (1996). Hazard ranking of landfills using fuzzy composite programming, *Journal of Environmental Engineering*, 122(4), 248–258.

73. Huang, G. H., Chen, Z., Tintowachwuthikul, P., & Chakma, A. (1999). Environmental risk assessment for underground storage tanks through an interval parameter fuzzy relation analysis approach, *Energy Sources*, 21, 75–99.

74. Van der Werf, H. M. G., & Zimmer, C. (1998). An indicator of pesticide environmental impact based on a fuzzy expert system, *Chemosphere*, 36(10), 2225–2249.

75. Mohamed, A. M. O., & Côté, K. (1999). Decision analysis of polluted sites—a fuzzy set approach, *Waste Management*, 19(7–8), 519–533.

76. Huang, C. F., & Yee, L. (1999). Estimating the relationship between isoseismal area and earthquake magnitude by a hybrid fuzzy-neural-network method, *Fuzzy Sets and Systems*, 107, 131–146.

77. Huang, C. F. (2002). An application of calculated fuzzy risk, *Information Sciences*, 142, 37–56.

78. Chen, Z., Huang, G. H., & Chakma, A. (2003). Hybrid fuzzy-stochastic modeling approach for assessing environmental risks at contaminated groundwater systems, *Journal of Environmental Engineering*, 129(1), 79–88. Doi: 10.1061(ASCE)0733-9372(2003)129:1(79).

79. Vemula, V. R. S., Mujumdar, P. P., & Ghosh, S. (2004). Risk evaluation in water quality management of a river system, *Journal of Water Resources Planning and Management*, 130(5), 411–423. Doi: 10.1061/(ASCE)0733-9496(2004)130:5(411).

80. Li, J., Liu, L., Huang, G., & Zeng, G. (2006). A fuzzy-set approach for addressing uncertainties in risk assessment of hydrocarbon-contaminated site, *Water, Air, and Soil Pollution*, 171, 5–18.

81. Li, J., Huang, G. H., Zeng, G., Maqsood, I., & Huang, Y. (2007). An integrated fuzzy-stochastic modeling approach for risk assessment of groundwater contamination, *Journal of Environmental Management*, 82(2), 173–188. Doi: 10.1016/j.jenvman.2005.12.018.

82. Sadiq, R., Kleiner, T., & Rajani, B. (2007). Water quality failures in distribution networks—Risk analysis using fuzzy logic and evidential reasoning, *Risk Analysis*, 27(5), 1381–1394. Doi: 10.1111/j.1539-6924.2007.00972.x.

83. Quelch, J., & Cameron, I. T. (1994). Uncertainty representation and propagation in quantified risk assessment using fuzzy sets, *Journal of Loss Prevention in the Process Industry*, 7(6), 463–483.

## Chapter 8

# Who Let the "Social Scientists" into the Lab?

**Eleonore Pauwels**
*Science and Technology Innovation Program, Woodrow Wilson International Center for Scholars, Ronald Reagan Building and International Trade Center, One Woodrow Wilson Plaza, 1300 Pennsylvania Avenue, Northwest #3, Washington, DC 20004-3027, USA*
eleonore.pauwels@wilsoncenter.org

## Introduction: Moving Away from 20th Century LCA

One of the reasons why new emerging technologies such as synthetic biology create growing public anxiety is they have now become part of the forces shaping the new era called "The Anthropocene" (Crutzen, 2002; IGBP, 2010; Robin & Steffen, 2007). In this situation, humans need to adapt to a new global environment that has never existed before, which is mostly created by us and with no clear references to the past to guide individual and collective action. Under the current trends of accelerated global change, where multiple socio-ecological

*Emerging Technologies: Socio-Behavioral Life Cycle Approaches*
Edited by Nora Savage, Michael Gorman, and Anita Street

ISBN 978-981-4411-00-4 (Hardcover), 978-981-4411-01-1 (eBook)
www.panstanford.com

constraints are increasingly intertwined (Rocktrom *et al.*, 2009), the development and diffusion of new technologies take on a completely new dimension: They not only serve to fulfill existing human needs and desires but become major forces shaping the dynamics of global systems in ways that in turn call for new and rapid forms of human adaptation.

In this context, traditional analyses focused on unveiling the potential health or environmental effects of new technologies, including life cycle analysis (LCA), are very much welcome, albeit insufficient to assess the broader aspects of sustainability of synthetic biology. What is at stake is our ability to understand the long-term viability of the new patterns of socio-ecological and socio-technical systems, which are likely to emerge from 21st century technological developments.

When it comes to LCA of emerging technologies, only limited research effort has been devoted to understanding the socio-economic impacts resulting from the introduction of new technologies within society. In contrast, considerably more attention has been paid to their impacts on the environment. Hybridization in the process of assessing ecological and socio-economic implications would benefit from being explored. What choice of indicators will make it possible to track important social values within a particular socio-ecological system? What are the social dimensions to be sustained? What tradeoffs have been made in the past between social and ecological sustainability and why? Traditional LCA is focused on tracking the environmental, and in some cases the economic, impacts of a technology. It is necessary to add the societal impact of emerging technologies to LCA to better understand the dynamics within the social formations driving sustainability transitions.

Given that we live in social systems that are organized, for the most part, around a plurality of values, research aimed at exploring social implications of emerging technologies should also involve anticipatory and participatory thinking. They should rely on experimental ways to integrate complex forms of knowledge assessment with more inclusive forms of stakeholders' engagement and citizen deliberation. One of these challenges is to improve the visualization of the socio-ecological and socio-technological choices we are faced with and to foster networks that bring practitioners together with scholars to promote the co-evolution of diverse forms of knowledge. A more inclusive form of LCA would be a useful tool in facing these challenges.

This chapter thus takes as its point of departure the assumption that LCA would gain from developing new ways of assessing innovations that are pluralist, inclusive of multiple disciplines, and, to a greater extent than at present, capable of implementing reflexive change and mutual learning, while maintaining a common focus on social robustness and sustainable, meaningful, and responsible developments. To this end, this chapter will explore the potential of using, upstream in LCA, collaborative epistemic networks such as interdisciplinary *trading zones* among scientists, engineers, ethicists, and social scientists.

A second assumption proposed in this chapter is that the articulation of plural techno-scientific visions and scenarios has to be embedded upstream in the LCA. First of all, interdisciplinary assessments of techno-scientific scenarios through trading zones would gain from occurring before the situation is locked in. If occurring upstream, these assessments will help understand the main factors, constraints, and drivers that are part of the technological innovation process. Moreover, trading zones organized upstream in LCA would contribute to addressing and translating technological uncertainties about potential risks to health and ecosystems in a way that can be treated by policy makers on a non-ambiguous mode. Finally, useful LCA will have to deal with the future in terms of both the evolution of technological innovation and its interaction with society, the environment, and other technologies. Again, trading zones would improve predictive accuracy if aggregating, upstream in the process, diverse perspectives from a variety of sources.

It is also proposed in this chapter that interdisciplinary trading zones upstream in LCA would be beneficial, especially if they allow the inclusion of downstream actors—such as end users and policy makers—in the technological innovation process. Contexts of innovation may be seen to consist of complex and intersecting knowledge and information-infused relations of professionals, technologies, laypersons, users, innovators, business and policy makers forming new constellations of collaboration, experimentation and reflection to meet societal challenges. An upstream dialogue between researchers and policy makers, for example, would help identify moments of safety or regulatory uncertainties in technological trajectories.

## State of the Art: Crossing the Line "In and Out" the Laboratory

Recent research in Science and Technology Studies (STS) has been conducted with a goal to develop new theoretical frameworks based on trading zones, shared expertise, moral imaginations, and epistemic cultures that show promise for understanding and facilitating interdisciplinary collaborations (Gorman, 2004; Gorman & Mehalik, Gorman *et al.*, 2004; 2002; Knorr-Cetina, 1999).

### Trading Zone, Interactional Expertise, and Cross-Field Collaborations

The metaphor of a "trading zone" was first developed by Peter Galison to explain how scientists and engineers from different disciplinary cultures manage to collaborate across apparently incommensurable paradigms (Galison, 1997). Through case studies in physics, Galison found that different epistemic communities had to first develop jargons, then pidgins, and finally full-scale creoles to be able to share perspectives across their own scientific paradigms. He noticed that despite coming from contrasting scientific paradigms, experts were able to develop communication processes that can be seen as a trading zone. In this trading zone, experts use what Bromme calls a group-specific language, which usually relies heavily on the development of metaphors as a bridge between different epistemic paradigms (Bromme, 2000). The concept of trading zones has been significant to better understand interdisciplinary collaborations. The "trade" metaphors adequately portray the way academic experts are increasingly used to meeting, exchanging ideas, learning from each other, and then returning to their epistemic community with concrete "goods" in the form of improved research practices.

In parallel with the reflection on trading zones, Collins and Evans (2002) have described the different levels of expertise that play a role in interdisciplinary collaborations: one of special interest to our research is the interactional form of expertise. The interactional expert corresponds to an agent who understands enough of the language and norms of the different epistemic cultures involved in the trading zone to facilitate the trade. For example, early in the development of MRI, surgeons interpreted as a lesion what an engineer would have recognized as an artifact of the way the device was being used; this breakdown in the creole between these communities was recognized and solved by an interactional expert who had a background in both physics and medicine (Baird & Cohen, 1999).

Trading zones might also incorporate collective overarching goals such as sustainability. Indeed, sustainability has emerged as the ascendant policy issue of the 21st century. While we continue to argue about the true definition of "sustainability"—particularly since it has become a fashionable buzzword for the policy community and related funding agencies—the challenge of converting our present socio-technical system to a "sustainable" system has developed as a new master narrative, inspiring policy discourse both in Europe and the United States. Trading zones that build around these collective overarching narratives need to apply what has been defined as "moral imagination" (Werhane, 1999). The concept of moral imagination assumes that human beings learn practical ethics from deep stories, collective archetypes, and unconscious, or often emotive dimensions of a problem or paradox that become models for ethical behaviors (Johnson, 1993). These stories are usually invisible, unquestioned, and progressively adopted as simple accounts of the reality (Sethi & Briggle, 2011). Moral imagination reasserts that these stories are contingent views that need to be confronted with the "epistemic other" (Wynne, 2009), when a serious dialogue about different world views is at stake.

This notion of moral imagination might also help shed light on some of the controversies around anticipatory governance. When it comes to anticipating human futures, some trends in STS have started to develop empirical methodologies that are capable of guiding decision making toward visualizations and framings that endorse multiple and varied values. These STS researchers have begun to reflect on models for engaging civil society actors and wider publics in the process of envisioning and assessing technological futures (Guston *et al.*, 2002; Sarewitz, 2005).

## Epistemic Cultures and Negative Knowledge

The framework of epistemic cultures analyzes the dynamics of knowledge production—their amalgams of arrangements and mechanisms bonded through affinity, necessity, and historical coincidence (Knorr-Cetina, 1999). This framework might be useful to explore the potential for new modes of cross-sector (public/private) and cross-disciplinary collaboration between the life sciences and social sciences to develop reflexivity in scientific practices. Synthetic biology, in particular, has witnessed the development of these "lab-scale interventions" (Fisher, 2007; Rabinow & Bennett, 2009). The rationale behind these collaborative ventures is to identify social and ethical controversies further upstream in the R&D process. These collaborations also promote bridges that enable the communication of ethical and regulatory insights from the social sciences and bioethics component back to the laboratory. At the same time, such collaborative practices would benefit from being anchored in trading zones involving outsiders to the lab and non-institutional networks such as DIYBio or private conglomerates.

Part of the dynamics within epistemic cultures, the notion of "negative knowledge" (Knorr-Cetina, 1999) and the problematization of the non-production of knowledge (Proctor & Schiebinger, 2008), might illuminate some of the controversies around the safety/societal implications of synthetic biology. The concept of negative knowledge seeks to analyze the limits of knowing, the mistakes we make in trying to know, the things that interfere with our knowing, what we are interested in, and what we do not really want to know. This notion of negative knowledge might give insight into the way research agendas are built and delimited, especially when it comes to research oriented toward solving societal ills.

In this contribution, we postulate that trading zones have the potential to improve LCA by integrating interdisciplinary assessments through gradual co-production of methodologies, analyses, and concepts. Trading zones could act as spaces for the articulation of plural scenarios about technological innovations and, *ipso facto*, promote the transmission of social, ethical, and regulatory controversies from social sciences to the lab and vice versa. Trading zones could also act as spaces for the articulation of plural forms of LCA—the LCA itself, not the technology, becomes the boundary object—with the view to help improve the process of engaging members of the trading zone in more mutually beneficial trades. The merit of using trading zones goes beyond learning about a certain technological pathway taken as the boundary object. The merit of the trading zone is in the "cohabitation" it creates, in the interactive and iterative dynamics it instills between technological actors.

## Case Study: Probing the Concept of Trading Zone within Synthetic Biology

To explore the potential of using trading zones upstream in LCA, this chapter will focus on a case study inspired from the emerging field of synthetic biology. The extensive use of engineering concepts and metaphors in the emergence of synthetic biology in public and political discourses portrays the field as one easy to grasp and, at the same time, a very appealing and promising endeavor. These mechanistic representations are anything but new in biotechnology and genetic engineering, where metaphors or images constructed to represent new processes, products, and their potential effects have widely adopted mechanistic models. A good example is the use of "chassis"—sometimes "safe chassis"—to define the basic functionalities of a bacterial genome on top of which forward-engineered biological systems can be implanted. One of the subsequent challenges lies in assessing the power and role of these metaphors within trading zones between not only scientific and engineering disciplines but also social sciences and humanities. A corollary is to work from the potential of cross-disciplinary collaborative methods to study and develop two-way communication processes about societal, ethical, safety, and regulatory issues from the social sciences to the laboratory and vice versa.

The above-mentioned case study explores how trading zones actually function as a space to produce reflexivity for the life sciences involved in synthetic biology and the social sciences interested in the related implications. The reflexive governance of science and technology seeks to understand the contingencies on and conditioning by its own representations and interventions in the process of social choice. Reflexive collaborative thinking on the social and normative dimensions of synthetic biology and its applications is furthermore of paramount importance for policy-making communities: What is the impact of the synthetic biology community and its engineering practices on social or biological systems? What roles do engineering metaphors play in the regulatory debate about safety? As an innovative and potentially disruptive technology, will the benefits of synthetic biology outweigh the costs?

## A Glance at Our Biotechnical Futures

Approximately 30 years ago, eminent scientists Waclaw Szybalski and Anna-Marie Skala pointed to new developments in science, which they suggested were giving birth to a "synthetic biology," a genetic frontier they placed beyond the mere analysis and description of existing genes to encompass the design of novel gene arrangements (Szybalski & Skala, 1978). Although Szybalski and Skala's 1978 assessment smacked then of prognostication, developments in genetics in the past two decades have made their vision a more concrete reality. In particular, advancements in DNA synthesis and sequencing have enabled the engineering of micro-organisms from discrete, or off-the-shelf, chemical parts, even allowing scientists to "design to specification" micro-organisms capable of performing novel functions. In 2006, for example, University of California–Berkeley researcher Jay Keasling and his colleagues at Amyris Biotechnologies succeeded in engineering a microbe to produce artemisinin, an ingredient in anti-malarial drugs. Another milestone was achieved in May 2010 by J. Craig Venter—an important figure in deciphering the human genome—and his research team when they successfully assembled the first synthetic bacterial genome and used it to take over a cell.

Thus, by promising a range of applications from bioenergy to biosensors, synthetic biology promises to have a transformative impact on the ways we engineer and manufacture biological matter. In brief, this new technology could turn specialized molecules into tiny, self-contained factories, creating cheap drugs and clean fuels. The following vision, described by Rob Carlson in an article in the *IEEE Spectrum* magazine, is a good example of the potential ontological changes we may be facing in this journey toward the molecular economy (Carlson, 2001:15):

> "In 50 years, you may be reading *IEEE Spectrum* on a leaf. The page will not actually look like a leaf, but it will be grown like a leaf. It will be designed for its function, and it will be alive. The leaf will be the product of intentional biological design and manufacturing. Rather than being constantly green, the cells on its surface will contain pigments controlled by the action of something akin to a nervous system. Like the skin of a cuttlefish, the cells will turn color to form words and images as directed by a connection to the Internet of the

day. Given the speed with which the cuttlefish changes its pigment, these pages may not change fast enough to display moving images, but they will be fine for the written word. Each page will be slightly thicker than the paper *Spectrum* is now printed on, making room for control elements (the nervous system) and circulation of nutrients. When a page ages, or is damaged, it will be easily recycled. It will be fueled by sugar and light. Many of the artifacts produced in 50 years and used in daily living will have a similar appearance and a similar origin. The consequences of mature biological design and manufacturing will be widespread, and will affect all aspects of the economy, including energy and resource usage, transportation, and labor."

This vision is simultaneously futuristic and foreseeable, reminding us that synthetic biology is ultimately part of a technological continuum anchored in the Enlightenment and constantly progressing through techno-scientific breakthroughs, such as recombinant DNA technologies.

Behind this impression of a continuum, however, there is something salient in the visions populating synthetic biology; through intentional biological design and manufacturing, engineered life forms—from engineered yeast to Venter's "synthetic cell"—are becoming "factories" on their own. In short, while laboratories have grown into "factories" through twentieth century's collective imaginaries, today synthetic biology design turns the living cell itself into a factory. To this effect, Galison (1999) remarkably analyzed how scientific practices and understandings have evolved through the nineteenth century from an Enlightenment culture seeking to unveil nature's true face to a regime of mechanical objectivity. Scientific practices have progressed from those of intervening genial individuals to ones at ease building and supervising precise machines. The following excerpt depicts the transformations occurring within the sanctuary of the laboratory (Galison, 1999:33–34):

"Many features of the laboratory and factory coincide; they are deeply linked, and often co-produced. One can point, for example, to worker discipline, centralized power sources, and architecture—as well as shared political economic ideals of maximizing work and minimizing waste. But for our purposes here, the key commonality is the joint fascination with the reduction of individual variability through the use of machines: the production of regularity as a positive virtue that

> was simultaneously moral and epistemic. It was here that the quieting of the will met the discipline and self-restraint of the factory.
>
> . . . .
>
> . . . Scientific laboratory workers had long taken on the mantle of self-disciplined supervisors of machine. When scientists announced with pride in objectivity that they would do nothing to impose individual variation on the regular, uniform, and reliable output of their machines, they were testifying not only to the power of science in industry, but to the conjoint understanding of laboratory and factory."

The vision of a future inhabited by "living factories" constitutes a significant and symbolic pace on the road to the molecular economy. It epitomizes and reinforces what some have called the production of "biovalue" within a "moral economy of hope" (Rose and Novas, 2005:442, 452):

> "Biology is no longer blind destiny, or even a foreseen but implacable fate. It is knowable, mutable, improvable, eminently manipulable. Of course, the other side of hope is undoubtedly anxiety, fear, and even dread at what one's biological future, or that of those one cares for, might hold. But whilst this may engender despair or fortitude, it frequently also generates a moral economy of hope, in which ignorance, resignation, and hopelessness in the face of the future is deprecated. This is simultaneously an economy in the more traditional sense, for the hope for the innovation that will treat or cure stimulates the circuits of investment and the creation of biovalue.
>
> . . . .
>
> . . . It also tries to encapsulate the ways in which life itself is increasingly locked into an economy for the generation of wealth, the production of health and vitality, and the creation of social norms and values."

This transition toward increasing reliance on the production of biovalue and the techno-scientific promises that surface in the aftermath presents a kaleidoscope of interesting epistemological and ontological claims. These claims predominantly rely on metaphors borrowed from engineering imaginaries and practices. For example, the influence of materials and computer engineering helps to explain synthetic biology's dominant vocabulary, with frequent references made to bricks, building blocks, fabs, open source, debugging, and plug-ins (Serrano, 2007). The extensive use of engineering concepts and metaphors in the emergence of synthetic biology portrays the field as one easy to grasp and, at the same time, a very appealing and promising endeavor (Specter, 2009). These mechanistic representations are anything but new in biotechnology and genetic engineering, where metaphors or images constructed to represent new processes, products, and their potential effects have widely adopted mechanistic models. Beyond the need to sketch the functioning of biological systems, these models also convey the implicit reassurance that these systems can be optimized and that they are reliable and under control; their behavior is predictable. This reassuring concept has also affected the design of regulation; mechanistic metaphors have been used as examples of mitigating uncertainties and managing safety aspects (OTA, 1989). Additionally, the effects of these images and metaphors are amplified by the fact that, as with most emerging sciences, the practitioners in charge of mapping synthetic biology are also concurrently inventing it (Sethi & Briggle, 2011).

There is no doubt that a lot of innovation will occur in the interstitial spaces between the disciplines involved in synthetic biology. But this emerging multidisciplinary smorgasbord will provide challenges in terms of the ability of new fields to regulate their own actions, anticipate unintended consequences, communicate effectively with each other and the public, and solve what some political scientists call "collective actions." There will likely be new challenges in managing ethical, social, and legal issues at the boundaries between disciplines. These emerging entanglements will give rise to questions and controversies—matters of concern—that we propose to highlight in the following point.

## Engineering Life or Engineering for Better Life?

Synthetic biology inspires controversy by claiming it can "engineer life." The claim is unprecedented among major scientific disciplines and suggests a commensurately unprecedented change to the way people understand and value nature. By virtue of its transformative objective, synthetic biology is at the forefront of what has been termed the "Molecular Economy" (Rejeski, 2010), as this integrative science borrows techniques and methodologies from a variety of disciplines, including genetics, molecular biology, information technology, and nanotechnology. Synthetic biology harnesses these fields in pursuit of the design and development of biological systems, frequently of high complexity, which do not occur in nature; the technology offers wide application in fields as diverse as energy, medicine, and materials engineering (Endy, 2005; Rodemeyer, 2009; Serrano, 2007). While promising great scientific innovation, particularly in the spaces between traditional disciplines, synthetic biology also presents serious challenges. The emerging technology's regulation and development, its ability, or lack thereof, to control for unintended consequences, and its very identity, especially its communication and relationship with non-scientific audiences, represent significant contemporary obstacles.

Paralleling the field's burgeoning development and applications—in particular at the interfaces between individual disciplines—new and still unimagined ethical, legal, and social dilemmas will likely emerge in the near future and significantly challenge the existing frameworks that guide scientific practice. While synthetic biology will no doubt blaze its own trail, the pathway it follows will likely resemble that of another pillar of the molecular economy: recombinant-DNA technology. Its emergence similarly sparked unimagined ethical, legal, and public health concerns, not all of which are yet resolved (Wright, 1986a,b).

Above all, in the scientific and public spheres, synthetic biology fits into a regime of innovation based on techno-scientific promises and therefore is epitomized through metaphors and narratives that involve the articulation of a vision (Wynne *et al.*, 2007). Often this articulation takes the form of hype. Vision and hype are both types of discourse that look toward the future. The vision of synthetic biologists is a future where humans engage in the large-scale design and creation of new life forms that are exquisitely tailored for human purposes. The genetic engineering of organisms and the extensive design and manufacture of living things from virtual genetic sequences blur the line between machine and organism, life and non-life, and the natural and the artificial and thus transform the relationship between humankind and nature in ways that are exciting to some people but troubling for others (Bedau *et al.*, 2009; Pauwels, 2009).

In the near future, there might be a need to explore the readiness of the engineering profession to address the ethical and social issues associated with our bio-technical futures. The possibility of error, human and otherwise, is the reason why it is important to consider history when we think about future technologies. How well have we managed the introduction of other technologies? Have we, as a society, learned anything?

Synthetic biology thus crosses important technological frontiers, like the boundary between science and engineering, and is part of what has been called the "New Biology" (NRC, 2009). Such a revolution in the life sciences, its nature and goals, preferably would require parallel adaptations in societal governance, but despite the efforts of visionary researchers to overcome the divisions between the two cultures of humanities and natural sciences (Jasanoff, 2004), the New Biology has been imagined mainly under the auspices of biologists, other natural scientists, mathematicians, and engineers. A comprehensive understanding of the epistemic, ontological, and normative changes induced by this New Biology paradigm would benefit from the involvement of researchers from humanities, including social sciences and bioethics.

## The "Two Cultures" Gap Revisited with Synthetic Biology

The successive reformulations of the nature and objectives of the life sciences—described in the above two sections—would gain from being accompanied by corresponding changes in the way synthetic biology is governed by and introduced into society. Thus far, policy responses to the development of new hybrid biological constructs have been quite limited in scope. Responses often take the form of creating ethics committees to study the implications of particular trajectories of research (EGE, 2009; PCSBI, 2010). This chapter argues in favor of a more comprehensive approach, addressing synthetic biology's full potential to influence human futures.

Too often, the public and policy debates surrounding synthetic biology have been narrowly focused around a utilitarian calculation of its technological benefits versus its potential regulatory risks. Although the technical aspects of synthetic biology policy are immensely important, spanning from controversies on ownership to socio-technical implications to biosecurity and biosafety concerns (nobody would like the re-engineered flu virus to mysteriously escape from the lab), fundamental questions about *what* applications of synthetic biology would advance societal goals and be considered sustainable are ignored and thus it limits the discussion to the opinions of a few technocratic elites. We need to think how to develop plural forms of LCA that are open enough to different epistemic cultures to be able to integrate these fundamental questions about how and when research in synthetic biology is being conducted to answer specific social and sustainability goals. This section continues with a recollection of attempts to consider different interests and concerns at the crossroad between science, society, and sustainability.

Some recent research initiatives have started to revisit what C. P. Snow called the "Two Cultures." Snow saw a growing divide between the cultures of the sciences and the humanities; a divide that continues to present an obstacle to responsible education and problem solving. The research initiatives discussed above promote different ways in which the cultures of science—far from standing apart from the rest of the academic disciplines—are in timely conversations with the cultures of the humanities, the social sciences, the arts, and the law.

One of these initiatives is called "lab-scale intervention." Nanotechnology—and to a limited extent, synthetic biology—has witnessed the development of these new modes of cross-disciplinary collaboration between natural sciences and humanities, which help develop reflective scientific practices (Fisher, 2007; Fisher & Schuurbiers, 2009; Fisher *et al.*, 2006). The rationale behind these collaborative ventures is to identify moments of ethical uncertainty and social controversies high upstream in the R&D process. These collaborations are also supposed to promote the transmission of ethical and regulatory issues from the social sciences and bioethics back to the laboratory. Encouragingly, recent studies show that it is possible to form an interdisciplinary trading zone in which a scientist and a humanist jointly explore a cutting-edge topic in nanotechnology (Fisher, 2007; Gorman *et al.*, 2004). Concretely, engineers and humanists become actively involved in the process of knowledge exchange, better described as "knowledge-trading," and consequently some engineers and humanists develop long-term interactions, building trust and enabling mutual learning by working together in hybrid collectives.

These long-term, cross-field collaborations are important for two reasons. On the one hand, such collaborations promote continuing communication "inside-out" the laboratory, which helps to ensure that there is mutual understanding and validation of the data produced. This refers to what Fortun and Fortun (2005) have described as the "ethnography of ethics"—assuming that reflexivity should also apply to social sciences—and "friendship with the sciences," which pictures a more positive collaborative engagement between lab scientists and embedded humanists. On the other hand, these collaborations sometimes function as forms of extended peer review, which favor cross-fertilization of knowledge (Funtowicz & Pereira, 2005).

In future, these collaborations of researchers from different disciplinary cultures could act as spaces for the articulation of plural narratives and metaphors that promote the transmission of scientific, ethical, and regulatory controversies from the social sciences to the lab and vice versa. This would function as a mirror or a "reflexivity tool" for the life sciences involved in synthetic biology design and the social sciences interested in the related implications. In a "knowledge society," this "reflexivity tool" could also be extended to the public sphere by including policymakers, NGOs, investors, and science journalists.

Ideally, such collaborative practices will require continual conversations with those outside the lab, including policy-making communities and non-institutional networks such as the Do-it-Yourself-Biology community and private conglomerates. Such an early dialogue between researchers and policy makers, for example, would help identify moments of safety or regulatory uncertainties in synthetic biology trajectories, or what Brian Wynne calls "epistemic other" (Wynne, 2009:13): "It is difference manifesting itself as an unknown set of realities, acting themselves as unknowns and beyond our control (but not beyond our responsibility), into a world we thought we controlled." Indeed, policy-making communities do not need only a clear perspective on the challenges posed by synthetic biology to ethics and politics, but they must promote, inside public policy communities, more reflexive thinking on the social and normative dimensions of synthetic biology design.

Though these cross-disciplinary attempts are still nascent, they already raise questions and require us to be critical: to what extent do these collaborative studies lead to better capacity to critically analyze the relevance of synthetic biology promises to societal goals? To what extent do they allow us to collectively experiment with possible alternatives within synthetic biology? To what extent will they succeed in developing co-production among multiple disciplines and perspectives from the outset as opposed to downstream reflection upon the ethical, legal, and social implications of synthetic biology?

## Experimental Trading Zone around a Biological Chassis

### Setting the scene of this trading zone

As a tentative answer to the above interrogations, this chapter introduces a few empirical reflections that arise from a six-month interdisciplinary study that included a limited group of about 15 experts in synthetic biology, technology assessment (including LCA) as well as regulators and members from civil society organizations. One goal of this working group was to bring together the technical, regulatory, policy and civil society worlds at an early stage of product development, in this case a genetically engineered arsenic test kit, so that concerns might be addressed upstream and during the development stage. A critical mass of the experts involved shared cross-field expertise, which allowed them to cross paradigms from life sciences, to social sciences and policy and, *ipso facto*, to shape and conceptualize the discussions along these boundaries. The working group gathered at several stages of the study in a kind of free-exchange—in which experts from different backgrounds jointly contributed to the eventual result. They also gathered around a common superordinate goal, which was flexible enough to be progressively re-framed. This superordinate goal is one of "sustainability;" the problem of arsenic contamination in Bangladesh and related efforts to create a low cost, low skill, accurate field test kit.

### Boundary object

The working group concentrated its effort on the LCA of what could be seen as a boundary object—an arsenic test kit based on a *rE. coli* chassis. The current design uses *E. coli*, lactose, and bromothymol blue where the presence of arsenic causes *E. coli* to break down the lactose and change the pH of the water, resulting in a change of the water color overnight from blue to yellow. The system uses parts that exist in nature; what is new is the combination of them together in the same strain of *E. coli*. The test device uses JM109, a commercially available strain of *E. coli*, engineered so that it cannot survive outside the laboratory and has been mutated to prevent the transfer of genes outside of itself. The device is still in the early stages of development, and the developers still envision many hurdles to overcome.

Here some more in-depth reflections on the "chassis" metaphor are needed, especially as it constitutes the boundary object of this trading zone. Every descriptive language is not only metaphoric and interpretative but also developed *ad hoc* to fulfill a certain agenda. As metaphorical and interpretative as it is, the biology as chassis metaphor is now central to the challenge of developing useful and supposedly safe forward-engineered bacteria.

The design of microorganisms endowed with a growing number of edited and chemically produced genes that rewire cells for given purposes constitutes one of the key objectives of contemporary synthetic biology. Yet, such genes may still constitute a very minor portion of the whole bacterial genome, which provides the genetic and metabolic background that allows the synthetic genes to do their program. The metaphor of a biological chassis fits to perfection here, as it evokes the image of an internal framework that supports a man-made object, for example, in a motor vehicle, the wheels and machinery. By the same token, the basic functionalities of a bacterial genome provide a chassis on top of which forward-engineered biological systems can be implanted. The very representation of a chassis suggests the autonomy of the peripheral, implanted genetic circuits in respect to the basic cell physiology and heralds the ease of engineering biological objects with predictable properties. But, beyond these supposedly reassuring claims, how should the broader assessment of this artifact be organized? Discussions within the working group emphasized the caution with which any safety assumptions should be taken into account. Discussants highlight several sources of uncertainty: (1) unlike modular parts used in civil or electrical engineering, biological components are extremely context-dependent; (2) they interact with metabolites and chemicals; (3) their combinations generate emerging properties; and, last but not least, (4) cells which harbor them do not remain the same size, but they grow and multiply. Several of these issues cannot be ignored and make the metaphor even more complex to grasp and delicate to use as a boundary object. In an interesting manner, the plasticity of the metaphor as a boundary object echoes the plasticity of the biological chassis itself.

The divergence between the engineering agenda and the actual properties of biological building blocks highlights the importance of picking the right metaphors and employing the right descriptive language when dealing with living systems. But safety implications are not the only reason to be interrogating the imaginaries that inspire the use of the "chassis" metaphor. In the "chassis" imaginary, the scientists appear as architect and builders, constructing the system in a way that will suit a certain functional purpose. They see themselves as modifying what is already in nature and, *ipso facto*, producing a form of biovalue that has a potential to be developed and commercialized. This way of imagining themselves as somewhat inventing something new rather than only discovering what is already in nature can be seen in connection with a patenting model in which the inventor has the right to appropriation and benefit. The scientific endeavor is considered as a genuine business: scientific activity is seen as fulfilling a particular social need and scientists are seen as entrepreneurs—crucial actors in the sustainability market of non-western countries.

Both the questions of safety and ownership as harbored by the boundary object—the *rE. coli* chassis —have implications for two remaining components of the trading zone, the moral imagination component and the use of interactional expertise.

## Moral imagination

In the course of the discussions about the biological chassis, the trading zone participants realized that "efficiency" is a key to understand the ways in which life is imagined in synthetic biology. Particularly in an extreme approach to synthetic biology, a good design is one that is efficient. In the field of synthetic biology, it is common to see a prominent number of applications, which are targeted to enhance "efficient" pathways to sustainability. For instance, designing algae that will efficiently capture $CO_2$ or designing bacteria that will detect arsenic in water are examples of bioremediation applications. Again, the idea is to design systems that are highly efficient and fully controllable. This efficiency approach to sustainability entails a radically mechanistic way of imagining nature. However, in many other cases, the idea of sustainability is built from differing ways of viewing nature: for instance, when nature is imagined as resisting standardization and being able to behave in unpredictable ways. We see how different ways of imagining the nature and complexity of living systems will influence different views on sustainability and the kinds of actions to be taken. This focus on the chassis metaphor and the related imaginaries helped open plural and broadly scoped discussions on values.

The focus on imaginaries may open up the possibility for two participants in a trading zone to realize that their values on issues like what constitutes sustainability are only "relative" perspectives, and that it is worth comparing these different perspectives (Gorman *et al.*, 2009). Moral imagination is not relativism—not all perspectives are equally true or false. But moral imagination is the capacity to step into another person's imaginary, which stimulates the search for new imaginaries, new solutions.

In contradiction with the idea that sustainability issues—whatever they are—can be solved by a "technological fix" and subsequent education of lay publics, this exchange between two trading zone participants is eloquent:

Participant 1: "There are different reasons for resistance and concern that isn't simply because of ignorance for some. Environmental groups who work with groups around the world have concerns for very real potential risks to environmental safety, public health safety and social economic issues. It's not enough to say 'well if you educate everyone then all those concerns are going to go away.' Looking at what we've already learned from the conversation earlier in the day and what more we will learn as the conversation goes on, that there are some very serious and difficult issues relating to bioremediation and others. The more we learn about the science and the technologies, the more we learn more about the risks that one might be able to identify. It's not enough to just say it's a matter of 'not knowing' or 'ignorance;' it's more about the knowledge that we've gained on the subject, the more concerns that can be raised. And it shouldn't just be dismissed as pure ignorance."

Participant 2: "Is there a slightly mistaken assumption that in order for one to engage in this conversation requires technical knowledge? To do the latter parts of the risk assessment, to answer some of the question in regards to risk and to understand those answers during the risk assessment may take technical knowledge but that, in theory, locks you into looking at only the technical question. If you look at the conversation of risk assessment as one part that sits within the bigger picture of risk management, which also can include the question of what are your concerns and what can be done to take care of those concerns. Not just saying 'this technical reason takes care of those technical concerns,' which can lead into big arguments about 'how can I trust that technical answer if I don't understand the technology.' Looking at risk management procedures, for people who don't necessarily have the technical background but do have concerns. Their questions are 'don't tell me all the reasons why I'm not going to get sick, tell me all the reasons, ways and what you are going to prevent that organism from getting out so I don't get sick.' This directs the developer to thinking about ways to carefully run field tests that don't necessarily require the technical knowledge of all the stakeholders. Creates an environment where the developer is taking these concerns into account and is addressing them and the other stakeholders can see how their concerns are being addressed. One should almost take a step back and look at the bigger picture, viewing the exercise from a risk management perspective when engaging in a multi stakeholder's assessment process."

The above excerpts show how important it is to problematize ignorance as a subject of socio-ecological and socio-technical inquiry. In a way, it is an invitation to problematize the sociological and historical roots of the dynamics that lead to non-production of knowledge about what and who we are supposed to sustain. However, this process of "opening-up," beyond the temptation to avoid complexity and rely on technological fix, is usually promoted by the presence of different forms of knowledge and expertise in the trading zone. We will address this crucial point under the terms of "interactional expertise."

## Interactional expertise

The discussion on moral imagination has begun to demonstrate the importance of introducing exogenous normativity into discourses of progress and the role marginalized and unconventional actors play in directing innovation. Such discussion might help to deconstruct the values, reasoning, and framings at stake in controversies over sustainability. In our experimental case study, this part of the discussion on moral imagination was systematically promoted by what can be seen as "interactional" expertise. A majority of the trading zone participants had enough knowledge and understanding of the language and norms of the different cultures involved in the zone to facilitate the trade. These agents were able to draw connections between the issues raised within the scientific and ecological discussions and also to put them into public policy and social perspectives. More specifically, they did not see the biological chassis only as a scientific invention that serves a specific social purpose but as a "living-tool" or better a "living-process" that would have to be introduced into a web of diverse practices: practices linked to human behaviors, cultures, and socio-ecological systems. Interestingly, what started as a "high-technological" conversation quickly focused on what participants decided to call "human practices." The free exchange of concerns below shows how the dynamic within the trading zone goes progressively from the boundary object—the biological chassis—to the same device being situated within a potential socio-ecological and socio-technical context. The exchange of concerns—the "trade"—acts as a tool producing an attitude of "reflexivity" among the participants.

Participant 1: "We need to understand human behavior in practice in order to evaluate and manage risks. For example, how users will dispose of the biosensor will have significant effects on risks. How human behavior is incorporated into risk models is a key piece of the discussion that is not well understood."

Participant 2: "How do you do human practice identification? You have to get the data from the community—talk to people or watch people. Does EPA have contractors to use to collect this data? There are government restrictions. Often to have to turn over to contractors, can be difficult—things get lost in the translation. EPA makes extensive use of contractors for many things."

Participant 3: "The way technology is deployed is important. What about regional and socio-economic value? People have to understand how to use the device. There is a need for a plan on how to educate the people and to popularize the device. There should be a standard disposal mechanism which should be transmitted to the people. Many people live in the rural areas. They may not know about the arsenic problem."

Participant 4: "Things to look at: survival of *E. coli*. What happens when it dies? What happens to the DNA? How does it die? This gets into values of stakeholders. Is there any harm? Issues of values become important."

Participant 5: "Who are the stakeholders and what are the public concerns relating to the chassis? What are the ecological ramifications for the microbial system? All of these factors and ideas of what is valuable differ."

Participant 6: "The stakeholder should be the planet, break it down into humanity and the ecosystem in which we live. Starting to think about it that way for now you can set that as a threshold, which in some ways makes things a bit more absolute, bound and clear then identifying all the subsets that you think of as a stakeholder. When it comes to micro-organisms that are likely to be released, either in the extremes of being released into the environment or if it's released into the body, then pretty soon after it's going to follow the same path as if it was released into the environment."

This shift in the discussion from the boundary object to the "stakeholders being the planet"— which targets potentially all socio-ecological systems where the device could be implanted—is nothing surprising. Only cross-field models of knowledge exchange—which draw upon and integrate empirical and theoretical elements from a variety of fields—may help in analyzing the interactions between multiple sets of actors as they interact in real-world entanglements. As argued by Voß and Kemp, "Considering the heterogeneity of the elements that play a part in sustainable development, effective problem treatment calls for the use of methods of integrated knowledge exchange that transcend the boundaries between disciplines and between science and society" (Voß *et al.*, 2006: 10-11). They also insist on the benefit of integrating "the tacit knowledge of societal actors"—which is "generated in interactive settings in which knowledge is co-produced by scientists and actors from respective fields of social practice." Concretely, citizen and groups concerned by the issues at stake get actively involved in the process of knowledge production, with the consequent result that some interactions between scientists and laypersons become permanent and build trust and mutual learning by working together in hybrid collectives.

## Critical evaluation of the functioning of this trading zone

We understand the experimental trading zone described above as an exercise of "anticipatory governance" (Barben *et al.*, 2007): assuming that promoting reflexivity at an early stage of scientific development may have an impact on scientific practices and policies. We depart from the idea that the scientific endeavor is organized in a way in which room for reflexivity (about, for example, ethical, societal, and socio-economic issues, issues "out of the lab") is rare. From the scientific method to the experimental situation, scientific practices rely on a systematic neglect of natural and social complexities and uncertainty. The aim of this experimental trading zone is, therefore, to make possible a rare space in which black boxes can be opened and scientists can concurrently take on their additional role as humans and citizens. We call it experimental for several reasons: first, integrating, upstream in the technological innovation process, such a diversity of fields and sectors is quite innovative; second, instead of being "static" and fixing opinions—a common aspect of stakeholders and public consultations that has been criticized by sociologists of science—trading zones allow for iterative mutual learning and negotiation. The data gathered along the way through laboratory research, reports, schemas, and presentations and shared between trading zone participants are subject to continual questioning, review, and improvement. Within the trading zone, scientists, social scientists, technology assessors, policy makers, and regulators have been encouraged to open the "black boxes" that lie along the path of development of the arsenic biosensor device. This trading zone was an opportunity to "travel out of the lab" and to imagine the possible applications and implications of the biosensor they were in the process of developing.

Trading zone, shared expertise, and moral imagination have only been combined in a few pilot projects called, as mentioned above, lab-scale interventions. This framework needs to be tested and modified in a variety of other interdisciplinary contexts. Eventually, through the below diagnosis, we aim at unveiling some of the difficulties, limitations, and room for improvement when it comes to using trading zone within LCA.

The main success of this trading zone is that it successfully constituted an exercise of anticipatory and inclusive governance. First of all, the exercise happened extremely early in the process—much earlier than usual procedures of assessment—before and during the development stage. While occurring upstream, the trading zone helped understand the main factors, constraints, and drivers that are part of the technological innovation process. Moreover, the trading zone contributed to addressing and translating technological uncertainties about potential risks to health and ecosystems in a way that can be discussed early and treated potentially by policy makers in an inclusive setting. In addition, the trading zone improved participants' predictive accuracy by aggregating, upstream in the process, diverse perspectives from a variety of sources. These perspectives were used to build scenarios to deal with the future both in terms of the evolution of the biosensor and its interaction with society, the environment, and other technologies. The fact that these interdisciplinary assessments of techno-scientific scenarios occurred before the situation was locked in was crucial. This indicates that scenarios might be an important complement to the use of metaphors and group-specific languages within trading zones.

A second success relates to the depth and scope of the discussions happening within the trading zone. The inputs, questions, and criticisms raised by the trading zone participants did not focus only on risks and environmental impacts but span a broad range of concerns from policy and regulatory concerns to issues linked to ownership as well as societal, socio-economic, and cultural implications. Table 8.1 below illustrates some of the concerns expressed by the trading zone participants at the end of the discussion process. These concerns, because they span from very different perspectives, reflect how the trading zone works as a "black boxes opening" tool, as a way of keeping a technological issue complex (Stirling, 2010).

**Table 8.1** Matters of concern within the trading zone

| |
|---|
| **Examples of ecological concerns** |
| "What will happen if the biological chassis enters the environment? Yes, every component already exists in nature, but what happens when they are combined? Will the new organism act the same? We don't know. How does the *E. coli* strain change in the field? Is there a risk of it getting stronger? What happens when it dies? When it comes to the synthetic organism and the receiving environment: Are you increasing the fitness of the organism? If it picks up another microbe's genes, can that increase fitness of the organism?" |
| **Examples of policy concerns** |
| "When to involve stakeholders in the LCA process? For what role? Are there faster avenues to provide assurances of environmental safety? What are the current regulatory standards and insurance standards?" "How will the public perceive the chassis? The public has a negative impression of GMOs and *E. coli*. The *rE. coli* chassis is based off of *E. coli*. How can you distinguish the difference between the two? Is it a question of missed perception?" |

**Examples of regulatory and socio-economic concerns**

"A lack of proof of harm does not equal safety. Where does burden of proof lie? We need to ask the right questions on risks and benefits. Consider the example of a Synbio company, created as private company in US then moved to Brazil. What of the regulatory challenges of US companies working in another country not covered by US regulations? There is the potential for export of harms. We need to treat the socioeconomic issues and the questions of liability. We should not look to the precedent of liability with GMOs and crops as a guide. There are issues in that case related to the fact that it is too hard for farmers to secure compensation for damages from contamination of their non-GMO crops."

Last but not the least, the trading zone gave an important space for the policy, regulatory, and civil society actors to voice their concerns and to intervene with feedbacks related to the development pathway of the biosensor. This reminds us that on the surface of this epistemic variety, a democratically committed knowledge society is supposed to have the scientific and political imaginations to work out how a plurality of social actors could share knowledge, practices, and experiences with diverse scientific, policy, and economic actors (Jasanoff, 2009). Beyond the distributed nature of expertise, what matters most is the consideration of the divergent socio-cultural contexts in which techno-scientific politics take place, *inter alia* the modes of knowledge making in the public sphere and the levels of accountability and trust in the knowledge produced (Jasanoff, 2005).

The critical reflections based on experimenting with trading zones invite us to provide a role for engagement of a more complex kind. In this case, scientists, engineers, and policy makers, sensitized through engagement to wider social imaginations, might decide for themselves to approach science and innovation differently. Interestingly, what interdisciplinary expertise help picture are hidden (1) *questions*, (2) *connections*, and (3) *suggestions* (Wilsdon, 2007): (1) questions might be about *what we don't know* and *how to learn what we should not do*; (2) connections might show the risky entanglements between public–private, global–local interests involved in techno-scientific promises; (3) suggestions might range from anecdotal evidence to alternative practice or alternative technology scenario. As explained by Stirling (2009) about current discourses on sustainability:

> *"Often, the position is expressed as if there were 'no alternatives.' The questions asked are thus typically restricted to 'yes or no?'; 'how much?'; 'how fast?' and 'who leads?' If we move instead to more plural understandings of progress, then the quality of debate—and of the ensuing choices—thereby stands to be enriched. Instead of fixating on some contingently-privileged path, we might ask deeper, more balanced and searching questions about 'which way?'; 'what alternatives?'; 'who says?' and 'why?' This is the essence of a normative, analytic, epistemic, ontological—and consequently intrinsically political—project of 'pluralising progress'."*

Stirling's reflection left us with a daunting yet challenging array of questions of how to open technological controversies to discussion. In this perspective, the work of Arie Rip brings remarkable insights into the value of (1) enabling future-oriented actions between actors who share an environment and (2) supporting them to create narratives about the potential resulting uncertainties and ambiguities (Rip, 2006). More importantly, Rip stresses the need for *diversity* as a source of renewal by creating grey zones and interstitial spaces in existing orders and institutions where dissenting imaginations might be voiced. This reminds us that the ultimate challenge is about reviving atmospheres of our democracies, which allow for the expression of dissenting imaginations. The ultimate challenge is avoiding "high-pace techno-scientific politics" to withdraw from the democratic scene (*learning is forgetting*) and to cultivate the ability of "making things public" and turning "matters of facts" into "matters of concern" (Latour, 2005). At the same time, decisions about technological innovation pathways have to be taken and cannot be indefinitely postponed. The main question thus "simply" becomes: how do we make better decisions by creating a space for mutual learning about technological pathways and their socio-economic and political ramifications. As suggested by this contribution, such a space could be built around an ongoing trading zone. In a vibrant call, Latour invites us to give a chance to what he names *cohabitation* (2005: 20):

> "Can we cohabitate with you? Is there a way for all of us to survive together while none of our contradictory claims, interests and passions can be eliminated?"

It is the unveiling of these improved forms of collaborations that I wished to explore in this contribution. And eventually the term "cohabitation" should be preferred to "collaboration." Indeed, the concept of collaboration itself is a matter to be discussed. Cross-field and cross-sector collaborations have too often been considered as "fusion"—where actors converge toward a premeditated vision or goal, suppressing *ipso facto* the room for a diversity of knowledges, practices, and experiences; too often, collaborations are experienced as an attempt to co-optation—meaning that the instrumental support of a field, such as ethics, philosophy, or sociology, is required to make up for an interdisciplinarity of "façade." The term "cohabitation" entails more: it presupposes that we leave enough room for different frameworks of thinking to seat together, exchange, and ultimately develop visions that are based on a true diversity of claims, knowledges, and imaginations.

This contribution constitutes a thought experiment (Gedanken experiment) around the concept of trading zone. Echoing Latour's quote, this contribution emphasize the importance of the notion of *cohabitation* between fields, sectors, cultures, and ways of approaching regimes of techno-scientific innovation. Ultimately, how do you create the infrastructures so that complex ways of thinking from different fields, sectors, and cultures can meet somewhere and learn from each other? How can we think about forms of "cohabitation," where researchers from different fields could reflect together on research design, research questions, and trajectories? Is it possible for different socio-technical imaginations to cohabit? What are the necessary conditions (institutional, epistemic, political, and cultural) to develop different forms and places for reflexivity in different contexts such as the educational systems, the policy systems, or the laboratories?

## Conclusion

The chapter aimed to fill critical knowledge gaps about the role of interdisciplinary interactions at the interface between life sciences and society. In this contribution, we postulate that trading zones have the potential to improve LCA by integrating interdisciplinary assessments through gradual co-production of methodologies, analyses, and concepts. Trading zones can act as spaces for the articulation of plural scenarios about technological innovations and, *ipso facto*, promote the transmission of social, ethical, and regulatory controversies from social sciences to the lab and vice versa. To explore the potential of using trading zones upstream in LCA, this chapter focused on a case study inspired by the emerging field of synthetic biology. The case study explored how trading zones actually function as a space to produce reflexivity for the life sciences involved in synthetic biology and the social sciences interested in the related implications. Beyond the academic community, the case study also included civil society actors and members of the policy-making communities, who needed not only a clear perspective on the challenges posed by synthetic biology to ethics and politics but also to promote inside public policy communities a more reflexive collaborative thinking on the social and normative dimensions of synthetic biology and its applications.

Current innovations, such as synthetic biology, have become part of what can be labeled *global change technologies*, thus engendering new socio-ecological structures that have never existed before on Earth. Innovations open up for possibilities to exploit or develop resources and to potentially improve quality of life in many ways, but they also lead to new structural disruptions, new forms of pollution, and in some cases, new large-scale irreversible transformations in global socio-ecological systems, which can be highly detrimental for sustainability. As such, this ambivalence cannot be handled using many of the assumptions, tools, and models, which have prevailed and have supported the growth of science, technology, and global markets till the present. *Global Change technologies* require carefully crafted appraisal procedures that promote anticipation and reflectivity in many of the prevailing practices in science and policy. Under the present predicament, we cannot afford to wait any longer for the unwanted—although not totally unexpected—negative consequences of global experiments derived from the careless implementation of new technologies. The effects can simply be too costly or just impossible to solve.

Immediate acceptability of synthetic biology cannot be taken for granted either. Corporations and public research institutes engaged in the development of these potentially global change technologies should also be very interested in promoting credible R&D governance arrangements capable of guaranteeing that these innovations fulfill the requirements, demands, and expectations of the societies in which such technologies are to be implemented—e.g., in terms of long-term sustainability. This would require moving away from the twentieth century LCA model and learning how to better place science and technology in an array of socio-ecological contexts, including a long-term, global context. The harnessing of synthetic biology for sustainability will only be possible if special attention is paid to supporting a myriad of agents located at the interface of science, policy, and the public, so they can actively contribute to the development of new appraisal procedures that link these scientific and technological innovations to the current sustainability challenges.

## References

Baird, D., & Cohen, M. (1999). Why Trade? *Perspectives on Science*, 7, 231–254.

Barben, D., Fisher, E., Selin, C., & Guston, D. (2007). Anticipatory governance of nanotechnology: foresight, engagement, and integration. In E. Hackett, O. Amsterdamska, M. Lynch, and J. Wajcman (Eds.), *The Handbook of Science and Technology Studies* (3rd ed.). Cambridge, Massachusetts: MIT Press.

Bedau, M., Parke, E. C., Tangen, U., & Hantsche-Tangen, B. (2009). Social and ethical checkpoints for bottom-up synthetic biology or protocells, *Systems and Synthetic Biology*, 3(1–4), 65–75.

Bromme, R. (2000). Beyond one's own perspective—the psychology of cognitive interdisciplinarity. In P. Weingart and N. Stehr (Eds.), *Practising Interdisciplinarity* (pp. 115–133). Toronto: University of Toronto Press.

Carlson, R. (2001). Open source biology and its impact on industry, *IEE Spectrum,* 38(5), 15–17.

Collins, H. M., & Evans, R. (2002). The third wave of science studies, *Social Studies of Science,* 32, 235–296.

Crutzen, P. J. (2002). Geology of mankind: The anthropocene, *Nature*, 415, 23.

EGE (European Group on Ethics in Science and New Technologies) (2009). Opinion n* 25—Ethics of Synthetic Biology. Available at http://ec.europa.eu/european_group_ethics/docs/opinion25_en.pdf.

Endy, D. (2005). Foundations for engineering biology, *Nature,* 438, 449–453.

Fisher, E. (2007). Ethnographic invention: Probing the capacity of laboratory decisions. In *NanoEthics* 1 (2), Springer. Doi: 10.1007/s11569-007-0016-5. Available at http://www.cspo.org/documents/Fisher_ProbingLabCapacity_Nanoethics-07.pdf.

Fisher, E., Mahajan, R., & Mitcham, C. (2006). Midstream modulation of technology: Governance from within, *Bulletin of Science Technology Society*, 26(6), pp. 485–496.

Fisher, E., & Schuubiers, D. (2009). Lab-scale intervention, *EMBO Reports Science & Society*, 10(5), 424–427.

Fortun, K., & Fortun, M. (2005). Scientific imaginaries and ethical plateaus in contemporary U.S. toxicology, *American Anthropologist*, 107(1), 43–54.

Funtowicz, S., & Pereira, A. G. (2005). Quality assurance by extended peer review: Tools to inform debates, dialogues and deliberations, *Theory and Praxis*, 2, 14. Jahrgang–Juni 2005, S. 74–79. Available at http://www.itas.fzk.de/tatup/052/gufu05a.htm.

Galison, P. (1997). *Image and Logic: A Material Culture of Microphysics.* Chicago: The University of Chicago Press.

Galison, P. (1999). Objectivity is romantic, *ACLS Occasional Paper—The Humanities and the Sciences*, 47, 15–43.

Gorman, M. E. (2004). Collaborating on convergent technologies—education and practice, *Annals of the New York Academy of Sciences*, 1013, 1–13.

Gorman, M. E., Groves, J. F., & Shrager, J. (2004). Societal dimensions of nanotechnology as a trading zone: results from a pilot project. In D. Baird, A. Nordmann, and J. Schummer (Eds.), *Discovering the Nanoscale.* Amsterdam: IOS Press.

Gorman, M. E., & Mehalik, M. M. (2002). Turning good into gold: a comparative study of two environmental invention networks, *Science, Technology & Human Values*, 27, 499–529.

Gorman, M., Werhane, P., & Swami, N. (2009). Moral imagination, trading zones, and the role of the ethicist in nanotechnology, *NanoEthics*, 3(3), 185–195. Doi:10.1007/s11569-009-0069-8.

Guston, D., David, H., & Sarewitz, D. (2002). Real-time technology assessment, *Technology in Society*, 24(1–2), 93–109.

IGBP (International Geosphere-Blosphere Programme) (2010). Annual Report 2010. The Royal Swedish Academy of Sciences, Stockholm. Retrieved from http://www.igbp.net/publications/annualreports/annualreports/annualreport2010.5.1b8ae20512db692f2a68000120 43.html .

Jasanoff, S. (2004). *States of Knowledge: The Co-Production of Science and Social Order*. London: Routledge Press.

Jasanoff, S. (2005). *Designs on Nature: Science and Democracy in Europe and the United States*. Princeton, NJ: Princeton University Press.

Jasanoff, S. (2009). Governing Innovation. Paper presented at the Symposium *Knowledge in Question*—A Symposium on Interrogating Knowledge and Questioning Science # 597. Retrieved from http://www.india-seminar.com/2009/597.htm

Johnson, M. (1993). Moral Imagination: Implications of Cognitive Science for Ethics. Chicago: University of Chicago Press.

Knorr-Cetina, K. (1999). *Epistemic Cultures: How the Sciences Make Knowledge*. Cambridge, MA: Harvard University Press.

Latour, B. (2005). From realpolitik to dingpolitik or how to make things public. In B. Latour and P. Weibel (Eds.), *Making Things Public—Atmospheres of Democracy* (pp. 14–43). Cambridge: MIT Press and Karlsruhe, Germany: ZKM.

NRC (National Research Council), Board on Life Sciences: Division on Earth and Life Studies. (2009). *A New Biology for the 21st Century*. Washington, DC: The National Academies Press.

OTA (1989). New Developments in Biotechnology: Patenting Life—Special Report, OTA-BA-370. Washington, DC: U.S. Government Printing Office.

Pauwels, E. (2009). Review of quantitative and qualitative studies on U.S. public perceptions of synthetic Biology, *Systems and Synthetic Biology*, 3(1–4), 37–46.

PCSBI (U.S. Presidential Commission for the Study of Bioethical Issues) (2010). *New Directions—The Ethics of Synthetic Biology and Emerging Technologies*, Washington, DC. Retrieved from http://bioethics.gov/cms/sites/default/files/PCSBI-Synthetic-Biology-Report-12.16.10_0.pdf

Proctor, R. N., & Schiebinger, L. (eds.) (2008). *Agnotology: The Making and Unmaking of Ignorance*. Palo Alto: Stanford University Press.

Rabinow, P., & Bennet, G. (2009). Synthetic Biology: Ethical Ramifications 2009, *Systems and Synthetic Biology*, 3, 99–108.

Rejeski, D. (2010). Managing the Molecular Economy, *The Environmental Forum*, 27(1), 30–34.

Rip, A. (2006). A co-evolutionary approach to reflexive governance—and its ironies. In J.-P Voß, D. Bauknecht, and R. Kemp (Eds.), *Reflexive Governance for Sustainable Development* (pp. 82–101). Cheltenham: Edward Elgar Publishing.

Robin, L., & Steffen, W. (2007). History for the Anthropocene. *History Compass*, 5(5), 1694–1719.

Rockström, J., Steffen, W., Noone, K., Persson, Å., Chapin, III, F. S., Lambin, E., *et al.* (2009). Planetary boundaries: exploring the safe operating space for humanity, *Ecology and Society*, 14(2), 32. Available at http://www.ecologyandsociety.org/vol14/iss2/art32/"32/.

Rodemeyer, M. (2009). New life, old bottles: Regulating first-generation products of synthetic biology. *Synbio 2*. Washington, DC: Woodrow Wilson International Center for Scholars. Retrieved from http://www.synbioproject.org/process/assets/files/6319/nano_synbio2_electronic_final.pdf.

Rose, N., & Novas, C. (2005). Biological Citizenship. In A. Ong and S. Collier (Eds.), *Global Assemblages: Technology, Politics and Ethics as Anthropological Problems* (pp. 439–463). Oxford: Blackwell.

Sarewitz, D. (2005). This won't hurt a bit: assessing and governing rapidly advancing technologies in a democracy. In M. Rodemeyer, D. Sarewitz, and J. Wilsdon (Eds.), *The Future of Technology Assessment*. Washington, DC: Woodrow Wilson International Center for Scholars.

Serrano, L. (2007). Synthetic biology: Promises and challenges, *Molecular Systems Biology*, 3(158), 1–5.

Sethi, L. M., & Briggle, A. (2011). Making stories visible: The task for bioethics commissions, *Science and Technology*, 27(2).

Specter, M. (2009). A Life of its Own—Where Will Synthetic Biology Lead us? *The New Yorker*, September 28, 2009.

Steffen, W., Crutzen, P. J., & McNeill, J. R. (2007). The anthropocene: are humans now overwhelming the great forces of nature? *AMBIO: A Journal of the Human Environment*, 36(8), 614–621.

Stirling, A. (2009). Direction, distribution and diversity! pluralising progress in innovation, sustainability and development. *STEPS Working Paper 32*. Brighton: STEPS Centre.

Stirling, A. (2010). Keep it complex, *Nature*, 468, 1029–1031. Available at http://steps-centre.org/wpsite/wp-content/uploads/stirling-paper-32.pdf

Szybalski, W., & Skala, A. (1978). Nobel prizes and restriction enzymes, *Gene*, 4, 181–182.

Voß, J.-P., Bauknecht, D., & Kemp, R. (2006). *Reflexive Governance for Sustainable Development*. Cheltenham: Edward Elgar Publishing.

Werhane, P. H. (1999). Justice and trust, *Journal of Business Ethics*, 21(2–3), 237–249.

Wilsdon, J. (2007). *Public Engagement in Science*, Report of the Science in Society Session, Portuguese Presidency Conference—The Future of Science and Technology in Europe, Lisbon, 8–10 October 2007, Luxembourg: European Commission, EUR 23334.

Wright, S. (1986a). Recombinant DNA and its social transformation 1972–1982, *Osiris*, 2nd Series, 2, 303–360.

Wright, S. (1986b). Molecular biology or molecular politics? The production of scientific consensus on the hazards of recombinant DNA technology, *Social Studies of Science*, 16, 593–620.

Wynne B., Callon, M., Eduarda Gonçalves, M., Jasanoff, S., Jespen, M., Joly, P.-B., *et al.* (2007). Taking European Knowledge Society Seriously, European Commission Report of the Independent Expert Group on Science and Governance, Luxembourg: European Commission, EUR 22700.

Wynne, B. (2009). Daring to Imagine. Paper presented at the Symposium *Knowledge in Question—A Symposium on Interrogating Knowledge and Questioning Science* # 597. Retrieved from http://www.india-seminar.com/2009/597/597_brian_wynne.htm.

Chapter 9

# What Are the Factors Affecting Anthropogenic Phosphorous Loss? A Case Study of Chaohu Watershed in Central China

**Zengwei Yuan, Huijun Wu, and Jun Bi**
*State Key Laboratory of Pollution Control and Resources Reuse, School of the Environment, Nanjing University, Nanjing 210023, PR China*
jbi@nju.edu.cn

## Introduction

With the rapid growth of China's economy, more and more nutrients such as nitrogen (N) and phosphorous (P) are utilized through human economic and consumption activities. Consequently, huge wastewater-containing nutrients are discharged into the water environment and have caused more and more serious eutrophication (Carpenter *et al.*, 1998; Chen *et al.*, 2008). Eutrophication is a process by which lakes becomes increasingly rich in plant biomass as a result of the enhanced input of plant nutrients mainly N and P (Golterman *et al.*, 1991). It has been shown to negatively affect water-use for fisheries, recreation, industry, agriculture, and drinking (Carpenter *et al.*, 1998). Eutrophication has also become

*Emerging Technologies: Socio-Behavioral Life Cycle Approaches*
Edited by Nora Savage, Michael Gorman, and Anita Street

ISBN 978-981-4411-00-4 (Hardcover), 978-981-4411-01-1 (eBook)
www.panstanford.com

one of the most important factors in the delay in Chinese economic growth, while at the same time actually pushing China away from its desired sustainable development goals (Liu & Qiu, 2007).

In order to mitigate eutrophication, the Chinese government has promoted a range of actions, including new policies and technologies (SEPA, 2002). These have made great progress and improved the environmental quality. While focused only on natural factors, such as lake sediment, sludge, and algae, influencing water quality, these new policies lack systematic and comprehensive analysis of the human activities, which primarily cause eutrophication (Huang *et al.*, 2008; Matsuoka *et al.*, 1986). Eutrophication has become increasingly serious in China, particularly in recent years (Li, 2006; Li *et al.*, 2004; Zhang & Liu, 2007; Zhao *et al.*, 2010).

For many lakes, excessive P, rather than N, is usually the limiting nutrient for growth of aquatic plants and thus the primary cause of eutrophication (Carpenter *et al.*, 1998; Drolc & Koncan, 2002). Meanwhile, as eutrophication mainly originates from human activities, such as the application of fertilizer and manure to crops, increased soil erosion and runoff from fields, feeding animals, and the discharge of urban and industrial wastes (Bennet *et al.*, 1999; Deunert *et al.*, 2003; Drolc and Koncan, 2002; Schroder *et al.*, 2004; Smil, 2000), reducing the impacts of these sources on surrounding water is critical to alleviating eutrophication of lakes around. For mitigating eutrophication, it is important to understand the driving forces of P loss to water in the socioeconomic system first. Therefore, tracing and identification of the anthropogenic P pathways or flows throughout socioeconomic ecosystem is necessary for eutrophication mitigation based on the concept of life cycle nutrient management (Kennedy *et al.*, 2007).

In order to identify and trace the sources of P loads into water environment, we launched a longtime research program to quantify the pathways of anthropogenic P and N flows in Taihu and Chaohu watershed. We conducted an analysis of anthropogenic P in Chaohu watershed in 2008 (Li *et al.*, 2010; Yuan *et al.*, 2011a; 2011b; 2011c), as a pilot, through which we concluded that the largest three contributors to P losses into surface water environment in the socioeconomic system are crop farming, large-scale breeding, and rural consumption. However, the question is how to identify and analyze the factors that cause the present situation of anthropogenic P loss.

Based on our previous field survey in Chaohu watershed, the aim of this study is to identify, categorize, and analyze the various drivers contributing to the anthropogenic P loss to water. For the presentation of Chaohu watershed, the strategies required to limit P drainage to the surrounding water are provided.

## Methodologies

### Study Area

Chaohu is located in the central of Anhui province between the Yangtze River and Huaihe River and is part of the Yangtze River watershed. As shown in Fig. 9.1, the watershed covers an area of about 13,000 km$^2$ of which the upper region, north of the Chaohu Gate, totals 9130 km$^2$ across six counties (Hefei, Feixi, Shucheng, Lujiang, Chaohu, and Feidong) and the downstream area, south of the Gate, covers 4365 km$^2$ across three counties (Wuwei, Hexian, and Hanshan) (Gao *et al.*, 2011). The total population of the watershed is approximately 9.636 million. As one of the five biggest lakes in China, Chaohu Lake is important for the local environment and

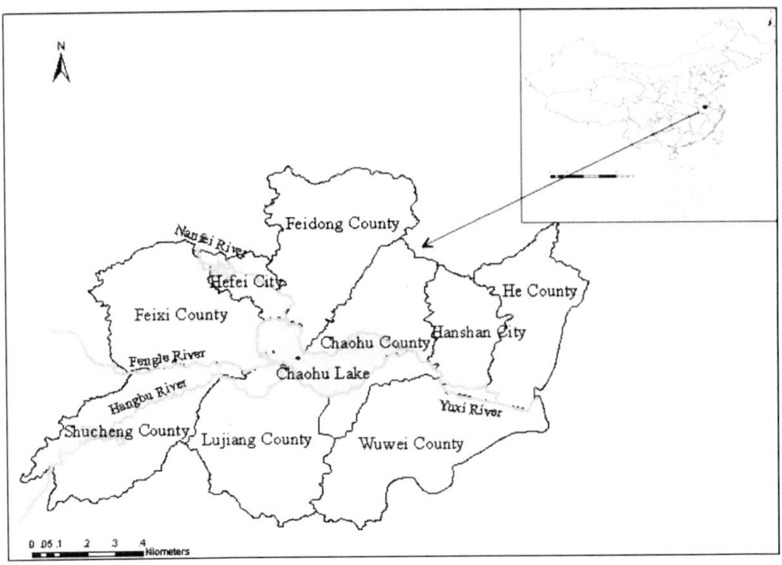


**Figure 9.1** The geographic location of Chaohu watershed.

society (Gao *et al.*, 2009). However, with rapid growth of economy and population, huge amount of pollutants are discharged into the lake and have been causing more and more serious water pollution especially eutrophication. Since the 1990s, eutrophication has become one of the most serious problems of Chaohu Lake (Wu *et al.*, 2009) and has created negative ecological, health, social, and economic effects on the lake and its use (Shang & Shang, 2005; Xu *et al.*, 2003; 2007).

## Analytical Framework

As two methods of quantifying the pathways of material flows, substance flow analysis (SFA) and life cycle assessment (LCA) are both based on the principle of mass balance. However, SFA is focused on one element throughout an ecosystem (entering, stock, and output), and LCA emphasizes all related materials to a specific object. Thus, SFA is usually used to establish a relationship among resources, economic and society development, and environment, and so it is often applied at macro level, such as county, region, and the world. LCA usually evaluates the environmental impacts of a human activity such as the production and use of a product by quantifying related materials throughout its entire life cycle, and so LCA is often applied at micro level, such as a firm. In recent years, some scholars have been applying LCA at the macro level to quantify the environmental impacts of economic development. However, the limitation of data at the macro level constrains its development. Usually, once a system boundary is defined, the internal processes and their sequences are decided at micro level, such as a firm. However, at macro level such as a socioeconomic ecosystem, the internal processes and their sequences are still in uncertainty and difficult to define with a commonly accepted way because they are usually changeable according to the different human activities. Therefore, it is difficult to establish a life cycle inventory at macro level, which is necessary for LCA. Furthermore, LCA aims at collecting all materials contributing to a specific environmental impact and so it needs great standard statistic system to acquire these data. Meantime, current statistic system is based on monetary indicators concerning little about materials and it brings a great challenge for conducting LCA at macro level.

In our previous analysis, we used SFA to quantify the pathway of anthropogenic P flows at the watershed level so as to identify sources of P loads. In this study, we developed an analytical framework to examine the factors causing P losses, which consists of sequential processes of drivers, environmental behaviors, and environmental performances (Fig. 9.2). The drivers could be grouped into three broad categories—the regulatory driver, the business driver, and the individual driver. On the basis of the actual situation of the study area and the data collected in the study, the key driving factors are selected for analyzing, and not all of the driving factors are included. Each of these three categories will be discussed as we identify the environmental behaviors that result in environmental performances representing with P losses (Li *et al.*, 2010; Yuan *et al.*, 2011a; 2011b; 2011c). Given that P has no stable atmospheric gas phase, unlike other nutrients contributing to eutrophication, such as nitrogen, the transfer of P could be learned as the state of aqueous phase in ecosystem (Filippelli, 2008). In the study, the environment is divided into surrounding water environment and soil environment because of our aim at tracing the sources of P flows contributing to eutrophication.

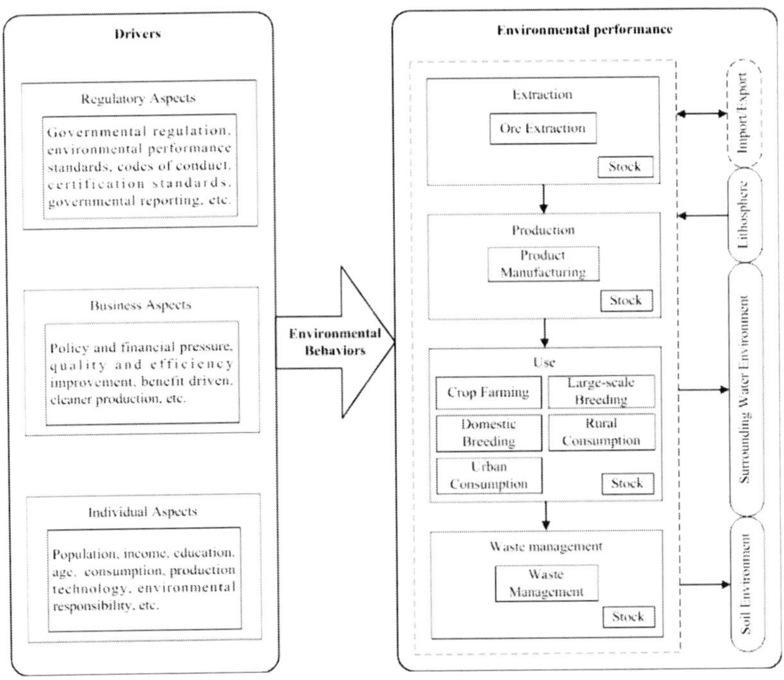


**Figure 9.2** Conceptual framework for driver identification of anthropogenic P loss into surface water environment.

## Data Collection

The approach for driver identification and environmental behavior analysis of environmental performance is rooted in the anthropogenic P flow analysis. Our data mainly come from statistical yearbooks, questionnaires, face-to-face interviews, and published literature. Statistical data is obtained mainly from the Statistical Yearbooks of Anhui, Hefei (including Hefei city, Feidong County, and Feixi County), Chaohu (including Chaohu city, Hanshan County, He County, Wuwei County, and Lujiang County), and Lu'an (including Shucheng County), which provide information about populations and products.

Two questionnaires are designed to evaluate the differences in consumption activities between rural and urban residents. One is for urban residents, focusing on the consumption and discharge of P-containing wastes. The other one targets rural farmers, which includes the input and output of farming and cultivation activities, in addition to the general contents of the first questionnaire. In Chaohu watershed, 2 cities, 16 towns within 7 counties, and 61 villages were selected on the basis of the economic level and geographical features, to distribute questionnaires and conduct interviews. A total of 1710 valid questionnaires were returned, of which 1078 were from rural residents and 632 from urban residents.

Moreover, to obtain objective data and meaningful information related to the study objective, we carried out the face-to-face interviews with officials from Anhui provincial, city, and county departments, residents and farmers, directors of industrial associations, such as Anhui Provincial Pesticide Association and Anhui Provincial Feeding Association, managers and staff of P-containing products manufacturing and consuming enterprises, retailers, and related experts. Some data about the P-containing rates in different substances were gathered from published literature.

## Results and Discussions

The anthropogenic P flow analysis indicates that the P losses to surrounding water in Chaohu watershed mainly originate from crop farming, large-scale breeding, and rural consumption, which accounted for 19621 t P, 12101 t P, and 2355 t P, separately. As the P loss is related to human activities, the status is explained from three aspects in the study: the regulatory aspect, the business aspect, and the individual aspect.

## Policy and Regulations

The governmental regulations have profoundly influenced anthropogenic P flows and increased the risk of P loss to water. Before 1978, with the slow economic development, nutrients tended to be self-sustained and are recycled in a closed loop in production and consumption, with livestock and human fed on the crops and crops fertilized by the human and livestock's manure. While the situation has changed since 1978 when the market-oriented economic form began and a so-called "household responsibility system" was established in rural areas, which linked remuneration to output and stimulated the structural changes in agricultural economy and rural society (Chen *et al.*, 2008; Tong *et al.*, 2003). Hence the traditional self-efficient production was substituted quickly. The spreading industrialization, rapid urbanization, intensive agriculture, and population growth resulted in a rapid increase of nutrients: People begun to buy more P-chemical products, such as fertilizers, pesticides, and feedstuffs, which generated much P discharged to water environment. As a result, the P cycle has been becoming unsustainable.

The intensive fertilizer use in plant production had an impact on the total P flow, causing 55.15% of P loss to water in Chaohu watershed in 2008. Meanwhile, as the nutrients are applied at rates that don't match crop intake, the runoff/erosion may be greatly increased (Sharpley & Smith, 1994), and the soil sediment accumulated in the crop farming was 18061 t P. With the unsustainable farming pattern, the most effective means of preventing P discharge into surface waters for the local government is to provide scientific farming technologies for farmers to apply fertilizers, encourage farmers to use non-/low-P-containing fertilizers, adjust planting structure, and develop the ecological agriculture.

On the other hand, livestock breeding is developing fast with the increasing demand of people. There are many breeding enterprises in Chaohu watershed especially in Feixi County. The government regulation concerning the issue has a negative impact on business competitiveness as it fosters an overly burdensome regulatory environment on smaller firms (Williamson *et al.*, 2006); many companies such as large-scale breeding enterprises discharge much wastewater to escape governmental regulation. The large amount of livestock manure has been far more than that can be taken up by local crops and arouses eutrophication of recipient water. As indicated above, the livestock manure from large-scale breeding is quite significant and requires further attention in the context of nutrient management as it contributes about 34.01% P loss to surrounding water, which is almost equivalent to that from crop farming. The government could force breeding enterprises to reduce the P losses and improve the environmental performance by enforcing the implementation of environmental regulations. Also, the government need provide breeding enterprises necessary finance and technologies for environmental management.

## Business Implication

Our field survey also shows that the enterprises will attempt to improve "business performance" rather than the "environment-protect performance" advocated by policy makers. The "business performance" criterion that drives the behavior of enterprises is far narrower than the "environment-protect performance" concept. "Business performance" is explained as focusing predominantly on cost reductions, efficiency, and satisfying customer needs, whereas "environment-protect performance" is more nebulous and diffuse, emphasizing benefits to a much wider group, including environment (Williamson *et al.*, 2006). This is because the enterprises can hardly see the benefit of taking a functional or environment-based approach, which they think to be inefficient. This is why enterprises play such an important role: It bridges the gap between the business' profit-orientated self-interest and the interests of environment. There are more than 3000 industrial enterprises in the watershed (Shang & Shang, 2005), and the agriculture holds the major property, while the P-chemical manufacturing industry is smaller. With wastes and wastewater drainage being subjected to stringent restrictions, the wastewater discharged to water from manufacturing is little, which is 99.09 t P. On the other hand, livestock production is developing rapidly with the increasing demand of people, which makes breeding wastes a serious problem. As a result, the breeding enterprises are the central concern with respect to water quality control.

At the same time, rural consumption and urban consumption contribute large P loss to water due to the consumption of many P-containing products. Though polices on P content in detergents is issued, while these policies are mainly optional and prices of P-free detergents are a bit higher than ordinary detergents, the residents tend to buy those with higher decontamination abilities and slightly lower prices. Therefore, the compulsory policies should be carried out to ban producing, selling, and advertising P-containing detergent and enhance public understanding on P-containing detergents polluting water. Moreover, it is found from our field survey that due to the backward environmental infrastructure in Chaohu watershed, especially in the countryside, rural wastewater and wastes have become the most significant pollution sources, which lead to the third largest pollution of eutrophication. So, it is suggested to develop sewage and solid-waste treatment facilities for towns, and even villages.

## Individual Implication

Humans consume a great many of P-containing products and produce a multitude of wastes, which consequently cause water pollution (Yuan *et al.*, 2011a). The crop food contributes the largest part of the consumption by both rural residents and urban residents, which indicates the consumption in the watershed is focused on the crop food other than the livestock food. Meanwhile, the livestock food consumed per urban resident is larger than rural resident because of the urban residents' higher living standard. However, rural and urban consumption drained a limited amount of P into the surrounding water system, accounting for 6.62% (2355 t) and 2.11% (750 t) of total P loss, respectively. It is because the rural population (6,766,000) is much larger than the urban population (2,870,000), and though P content of detergents is controlled, there is still a small quantity of P in some detergents (information gathered through investigation in the watershed, especially in the rural area). In addition, rural population consumes the commodities, including meats, eggs, milk, crops, and coal, more than the urban residents, which results in the much more P loss to surrounding water in rural consumption subsystem than that in urban consumption subsystem.

It is indicated from Fig. 9.3a,b that the population and income level are consistent with P loss to water of consumption sytem. Generally, the rural and urban areas with more population would generate much more P loss to water. It is noticed that the two counties of Wuwei and Lujiang with the largest populations among the nine cities/counties in the watershed have discharged the larger

P to the water than those of others. Similarly, among the urban areas, Hefei city, with the largest urban population, contributes the largest P loss to water.

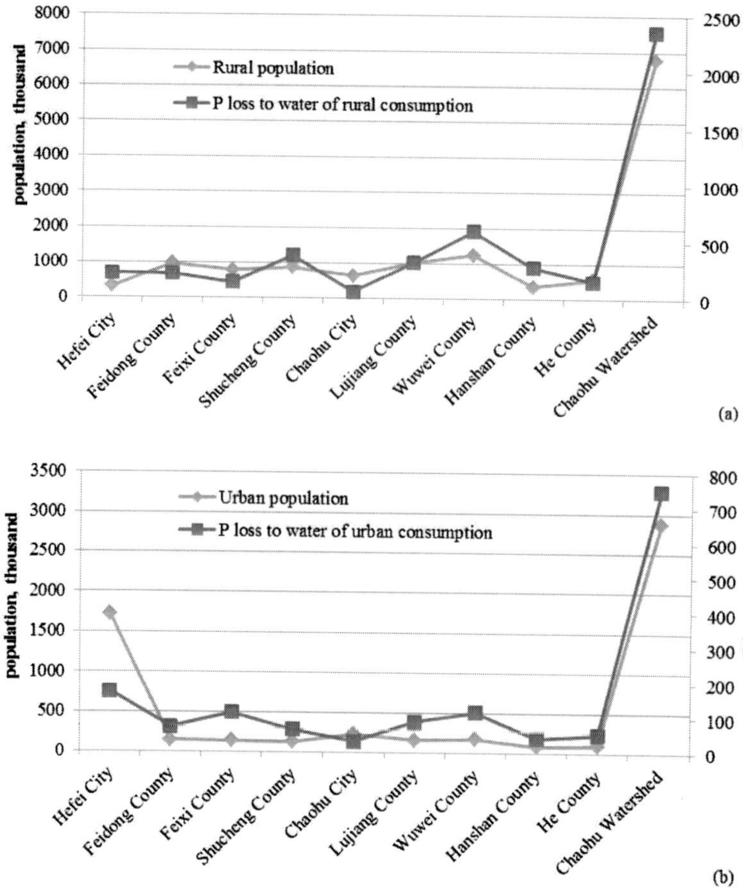


Figure 9.3 (a) The rural population and P loss to water of rural consumption in nine cities/counties of Chaohu watershed in 2008. (b) The urban population and P loss to water of urban consumption in nine cities/counties of Chaohu watershed in 2008.

There is a close relationship between the residents' income levels and the consumption volumes of P-containing commodities, which connects the income with the P loss to water, seen in Fig. 9.4a,b. The higher the income level, the larger P loss to water. Especially in Hefei city, as it is the capital of Anhui province and the economic center of Chaohu watershed, the income per resident in the city is the highest of the watershed, with more P-containing products consumed and wastes discharged. It is noted that with the improvement of rural life quality, the consumption of P-containing products and discharging of wastes are almost increasing in the same way as that in urban areas. The survey has demonstrated the relation of P loss to water with the income per resident.

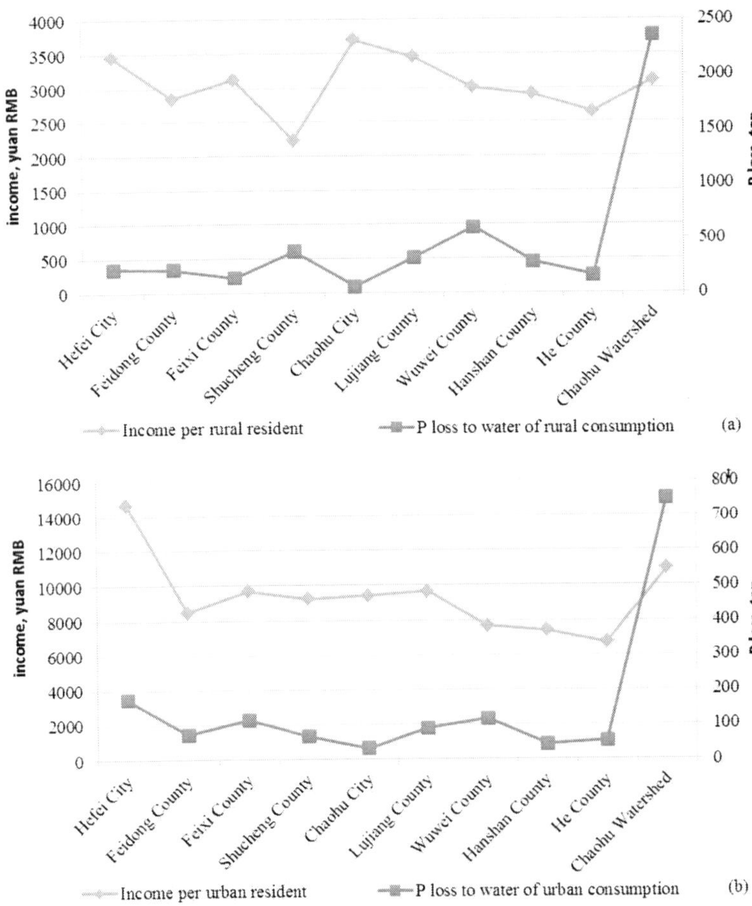


**Figure 9.4** (a) The income per rural resident and P loss to water of rural consumption in nine cities/counties of Chaohu watershed in 2008. (b) The income per urban resident and P loss to water of rural consumption in nine cities/counties of Chaohu watershed in 2008.

It needs to be noticed that elderly people and children usually stay at home and young and middle-aged people often work outside. This situation makes the residents' living structure simpler and the consumption and discharge smaller.

Actually, with respect to population size, income level, and age structure, which influence the lifestyles, changing consumer behavior, can significantly reduce P consumption. One of the biggest challenges is the adjustment of the lifestyle in order to make a safe and sound environment. Public participation on environmental aspects is of great importance (Shang & Shang, 2005).

From survey, we found that the residents' attitude toward environment in the watershed is not positive and a large part of their excrement and urine is discharged in the environment without recycling or disposal. In fact, the attitude toward recycling is strongly and significantly correlated with perceived control and situational factors, suggesting that the attitude is strongly associated with acquired knowledge (Tonglet *et al.*, 2004). So the local governments need to improve residents' environmental consciousness by enforcing environmental education.

Individual ecological responsibility is also a driving force that stems from the concern that an individual has for his/her social obligations. A salient feature of individual ecological responsibility is a concern for the social good. It is assumed that individuals might be motivated by a sense of obligation or responsibility rather than out of self-interest (L'Etang, 1995).

## Methodological Limitations

Improving environmental performance has become a more prominent requirement in all fields. The identification and analysis of its drivers and behaviors is made within the framework of anthropogenic P loss, while it is still featured by many uncertainties.

The first issue to be discussed is the matter of model structure. As P flows are seldom subject to full monitoring and documentation, especially in developing countries like China (Chen *et al.*, 2008), some of the data in the model are obtained from the literature and experts' experience. Where this is not applicable, the model structure would be simplified. It is assumed that P loss to water has a linear relationship with surface P balance in the model. However, the relationship is much more complicated, considering the natural factors such as wind erosion.

It must also be kept in mind that the system boundaries in this study are not sharp.

First, SFA only considers a single material and the problems related to this material (Udo de Haes *et al.*, 1997). Hence, other materials, such as energy and monetary flows, are not considered. Second, the study is based on a static analytical model and the data and results could only reflect the situation of P flows for the 2008 period. Therefore, the temporal changes in P flows could not be evaluated. For a comprehensive understanding of P stock variation in the region, long-term analyses are required. Third, the anthropogenic P flows could be subjected to further study and analyzed at a range of different scales. For example, different results would be expected on the basis of whether lake watershed or larger area is selected for study. Hence future work of greater accuracy should consider not only local areas but also national or even global areas.

Furthermore, there was a degree of uncertainty in the data used in this study, particularly with respect to the P composition rates of different materials. Nevertheless, the study sought to minimize these uncertainties by using and comparing an extensive range of reference sources, including published literature, investigations, and interviews. Considering the inadequate data sources in the study, it is difficult to evaluate the level of uncertainty in this approach. Further study on data uncertainty may be focused on identifying more detailed uncertainty intervals for assessing different sources of information.

Finally, as human activities are complicated, the driving forces of the anthropogenic P loss to water may be various. Considering the situation of the socioeconomic system of the watershed and the data collected, the study only identifies and analyzes several potential factors affecting the environmental behaviors greatly. Much of the analysis utilized in the study is qualitative, and the accurate identification of various drivers that result in the anthropogenic P loss needs to be improved. Hence further research would be valuable by building a quantitative relationship between drivers and environmental performance expressed with specific indicators.

## Conclusions

The problem of eutrophication exacerbated by human activities presents a serious challenge to sustain the development of regional economy and society. SFA model can be used to identify the most contributors of P loss by quantifying the pathway of P flows throughout a socioeconomic ecosystem. It is more convenient than LCA in tracing the pathway of a specific substance at macro level, such a socioeconomic ecosystem.

Policy is the major determinant of human recycling behavior. The spreading industrialization, rapid urbanization, intensive agriculture, and population growth are bringing about the rapid increase of P loss to water. Governments need to support the sustainable actions for the business and the public by providing preferential policies and provide farmers with technologies to improve the unsustainable farming pattern. Also, they need to encourage large-scale breeding enterprises to reuse, recycle, and dispose the wastes so as to reduce P loss. Furthermore, environmental regulation implementation needs to be enforced to reduce or eliminate the illegal discharge of wastewater, hazardous wastes, and untreated air emissions from enterprises.

Human environmental behavior is affected by the consumption style and the environmental consciousness of residents. Thus, it is important to improve pubic environmental consciousness by enforcing environmental education so as to change human environmental behavior and consumption style. Further research would be valuable for policy-makers by examining the internal and external drivers of environmental performance expressed with specific indicators, and evaluating impacts of greening public procurement in different sectors.

## Acknowledgement

The research was financially supported by the Natural Science Foundation of China (41222012) and the Fundamental Research Funds for Central Universities.

## References


Bennett, E. M., Reed-Andersen, T., Houser, J.N., Gabriel, J.R., & Carpenter, S.R (1999). A phosphorus budget for the Lake Mendota watershed. *Ecosystems*, 2(1), 69–75.

Carpenter, S. R., Caraco, N. F., Correll, D. L., Howarth, R. W., Sharpley, A. N., & Smith V. H. (1998). Nonpoint pollution of surface waters with phosphorus and nitrogen. *Ecological Application*, 8(3), 559–568.

Chen, M., Chen, J., & Sun, F. (2008). Agricultural phosphorus flow and its environmental impacts in China. *Science of the Total Environment*, 405(1–3), 140–152.

Deunert., F., Lennartz, B., & Tiemeyer, B. (2003). Legislative effects on the development of surface water quality in rural areas in Northern Germany. *Journal of Cleaner Production*, 15(16), 1507–1513.

Drolc, A., & Koncan J. Z. (2002). Estimation of sources of total phosphorus in a river basin and assessment of alternatives for river pollution reduction. *Environment International*, 28(5), 393–400.

Filippelli, G. M. (2008). The global phosphorus cycle: Past, present, and future. *Elements*, 4, 89–95.

Gao, C., Wang, X. Y., Jiang, T., & Jin, G. J. (2009). Spatial distribution of archaeological sites in lakeshore of Chaohu Lake in China based on GIS. *Chinese Geographical Science*, 19(4), 333–340.

Gao, F., Deng, J. C., Xu, Z. B, Ning, Y., Yin, H. B., & Gao, J. F. (2011). Ecological characteristics of macrobenthic communities in the Chaohu Lake Basin and their relationship with environmental factors. *Journal of Animal and Veterinary Advances*, 10(5), 627–634.

Huang, Q., Rozelle, R., Wang, J., & Huang, J. (2008). Water management institutional reform: A representative look at northern China. *Agricultural Water Management*, 96, 215–225.

Kennedy, C., Cuddihy, J., & Engel-Yan, J. (2007). The changing metabolism of cities. *Journal of Industrial Ecology*, 11(7), 43–59.

L'Etang, J. (1995). Ethical corporate social responsibility: A framework for managers, *Journal of Business Ethics*, 14, 125–132.

Li, R. Z., Wang, J. Q., & Qian, J. Z. (2004). Countermeasures of non-point nutrient control of Chaohu Lake Basin. *Journal of Soil and Water Conservation*, 18, 119–121 (in Chinese with English abstract).

Li, S. J. (2006). An approach to accelerating innovative development of the lake science in China. *Bulletin of Chinese Academy of Sciences*, 21(5), 399–405 (in Chinese).

Li, S. S., Yuan, Z. W., Bi, J., & Wu, H. J. (2010). Anthropogenic phosphorus flow analysis of Hefei City, China. *Science of Total Environment*, 408(23), 5715–5722.

Liu, W., and Qiu, R. L. (2007). Mini-review: Water eutrophication in China and the combating strategies. *Journal of Chemical Technology and Biotechnology*, 82, 781–786.

Matsuoka, Y., Goda, T., & Naito, M. (1986). An eutrophication model of Lake Kasumigaura. *Ecological Modelling*, 31, 201–219.

Schroder, J. J., Scholefield, D., Cabral, F., & Hofman, G. (2004). The effects of nutrient losses from agriculture on ground and surface water quality: The position of science in developing indicators for regulation. *Environmental Science and Policy*, 7(1), 15–23.

SEPA (State Environmental Protection Administration). (2002). *Study on Lake Eutrophication and its Countermeasures in China*. Beijing: China Environmental Sciences Press (in Chinese).

Shang, G. P., & Shang, J. C. (2005). Causes and control countermeasures of eutrophication in Chaohu Lake, China. *Chinese Geographical Science*, 15(4), 345–354.

Sharpley, A. N., & Smith, S. J. (1994). Wheat tillage and water-quality in the southern plains. *Soil & Tillage Research*, 30, 33–48.

Smil, V. (2000). Phosphorus in the environment-natural flows and human interferences. *Annual Review of Energy and the Environment*, 25(1), 53–88.

Tong, C. L., Hall, C. A. S., & Wang, H. Q. (2003). Land use change in rice, wheat and maize production in China (1961–1998). *Agriculture, Ecosystems and Environment*, 95(2–3), 523–536.

Tonglet, M., Phillips, P. S., & Bates, M. P. (2004). Determining the drivers for householder pro-environmental behavior. *Resources, Conservation and Recycling*, 42, 27–48.

Udo de Haes, H. A., van der Voet, E., & Kleijn, R. (1997). Substance flow analysis (SFA), an analytical tool for integrated chain management. In: Bringezu, S., Fischer-Kowalski, M., Kleijn, R., and Palm, V. (Eds.), *Regional and National Material Flow Accounting: From Paradigm to Practice of Sustainability* (34–42). *Proceedings of the 1st ConAccount Workshop, Leiden, Netherlands*. Wuppertal: Wuppertal Institute.

Williamson, D., Lynch-Wood, G., & Ramsay, J. (2006). Drivers of environmental behaviour in manufacturing SMEs and the implications for CSR. *Journal of Business Ethics*, 67(3), 317–330.

Wu, M., Zhang, W., Wang, X. J., & Luo, D. G. (2009). Application of MODIS satellite data in monitoring water quality parameters of Chaohu Lake in China. *Environmental Monitoring and Assessment*, 148(1–4), 255–264.

Xu, F. L., Tao, S., Dawson, R. W., & Xu, Z. R. (2003). The distributions and effects of nutrients in the sediments of a shallow eutrophic Chinese lake. *Hydrobiologia*, 429, 85–93.

Xu, J., Zhang, M., & Xie, P. (2007). Stable carbon isotope variations in surface bloom scum and subsurface seston among shallow eutrophic lakes. *Harmful Algae*, 6(5), 679–685.

Yuan, Z. W., Liu, X., Wu, H. J., Zhang, L., & Bi, J. (2011a). Anthropogenic phosphorus flow analysis of Lujiang County, Anhui Province, Central China. *Ecological Modeling*, 222(8), 1534–1543.

Yuan, Z. W., Shi, J. K., Wu, H. J., Zhang, L., & Bi, J. (2011b) Understanding the anthropogenic phosphorus pathway with substance flow analysis at the city level. *Journal of Environmental Management*, 92(8), 2021–2028.

Yuan, Z. W., Sun, L., Bi, J., & Wu, H. J. (2011c) The phosphorus flow analysis of Shucheng city, China. *Ecological Application*, 21(7), 2822–2832.

Zhang, Z., & Liu, G. M. (2007). The progress and thinking of lake eutrophication treatment in China at present. *Industrial Safety and Environmental Protection*, 33, 50–52 (in Chinese with English Abstract).

Zhao, Y. H., Deng, X. Z., Zhan, J. Y., Xi, B. D., & Lu, Q. (2010). Progress on preventing and controlling strategies of lake eutrophication in China. *Environmental Science and Technology*, 33, 92–98 (in Chinese with English Abstract).

Chapter 10

# Life Cycle Assessment and the U.S. Policy-Making Context

**Steven A. Cohen**
*The Earth Institute, 2910 Broadway, MC 3277, Hogan Hall, Level A, New York, NY, USA*
sc32@columbia.edu

## Introduction

Environmental policy makers use life cycle assessment (LCA) to reduce scientific uncertainties about the environmental impact of technologies and products. However, the structure of the U.S. policy-making system often acts as a roadblock to the use of LCA outputs in decision making. Over the past two decades, life cycle assessment has had limited impact in determining U.S. environmental policy, either as a specific quantitative indicator or as a general outlook toward the environment. In this chapter, I discuss specific characteristics of the U.S. political system, which often make uncertainty-reducing assessments like life cycle assessment fail in a decision-making context: the incremental nature of the U.S. policy-making system; the

*Emerging Technologies: Socio-Behavioral Life Cycle Approaches*
Edited by Nora Savage, Michael Gorman, and Anita Street

ISBN 978-981-4411-00-4 (Hardcover), 978-981-4411-01-1 (eBook)
www.panstanford.com

place-based nature of environmental politics in the United States; and the uncertainties associated with most environmental and scientific challenges. These three essential components of environmental policy-making in the United States pose inherent challenges to the use of life cycle assessment perspectives in policy debates.

We cannot look at life cycle assessment or any other environmental decision tool in a vacuum. The effectiveness of such tools hinges on the way they interact with the political context in which environmental decisions are made. To maximize the practical relevance of tools such as LCA, we must tailor them to suit the real world of environmental politics. In this chapter, I attempt to explain the limited scope of LCA in U.S. policy-making by taking a look at the unique benefits of LCA outputs and then viewing these against the backdrop of the U.S. environmental policy-making context. In doing so, I first present a historical overview of the relationship between life cycle assessments and policy-making; next I explain the incremental nature of U.S. policy-making, the role of uncertainty in the policy-making, and the place-based nature of U.S. policy-making. The end of this chapter ties its environmental policy-making lessons together with a look at the role of life cycle assessment in the politics of climate change in the United States.

The purpose of this volume is to consider how we might increase the scope and effectiveness of life cycle assessments. In this volume, many authors have indicated some of the technical areas where LCA methodology should be refined. I argue that beyond these technical alterations, improving the "fit" of life cycle assessment outputs into the U.S. policy-making context is an important way to make LCA more effective. Without proper appreciation and consideration of the constraints posed by the relevant political context, the practitioners will continue to experience frustration in achieving the necessary political relevance of LCA—even as the methodology evolves. Methodology and studies should be designed to not only enhance the scientific and the technical validity of an LCA but also to maximize their political efficacy and the probability that LCA outputs actually inform policy debates—both goals are equally important and should be pursued simultaneously.

## LCA as a Policy-Making Tool

Policy makers often have to make decisions with limited knowledge of the environmental impacts of the things they are making decisions about. Human activities have many consequences that fall under the umbrella of environmental impacts, and it is almost impossible to deduce the overall preferability of one set of options over another. Policy makers concerned with environmental protection must therefore rely upon scientific knowledge produced by tools such as life cycle assessment, impact assessment, cost-benefit analysis, and risk assessment to minimize uncertainty about products and processes and to inform decisions that allow them to implement the least environmentally harmful policies. The fundamental feature that sets life cycle assessment apart from other techniques is that LCA tracks the environmental effects of a product or a process from cradle—the resources used to create a product—to grave—the waste that results from its disposal, and including the resources used during its lifespan (Finnveden, 2000; Ross & Evans, 2002; US EPA, 2006). The output of LCA studies is an assessment of the environmental impact of a product or a service system over many different categories relative to the benefit provided by that system (Schenck, 2009). Product or service systems are different aspects of human activity, such as generating hydroelectric power, transporting goods, disposing waste, etc. We might want to know, given two options for creating the same benefit and a set of criteria to consider, which option we should select. Therefore, LCA outputs can be instrumental in evaluating different environmental tradeoffs between alternate means of providing a similar good or service. Furthermore, LCA outputs are expressed numerically, and this provides ease of interpretation for policy makers interested in comparing these alternatives (Cowell *et al.*, 2002, 881). Quantitative outputs allow policy makers to analyze the environmental damage of a product or a service simply and can allow effective and targeted decisions based on environmental criteria. This comprehensibility has proven to be one of the major benefits of LCA in political settings.

For example, once a particular substance has been targeted as harmful or hazardous to human health, LCAs can be used to adopt policies that minimize the use of that substance in a given process or product (Curran, 1997, 42). An example of this can be seen in

California, where LCA analyses have been used to minimize the leaching of heavy metals from shredder residue in landfills (California Environmental Protection Agency, 2007). Shredder residue consists of glass, plastics, carpet, and dirt, which remain after discarded automobiles and appliances are shredded and the valuable metals are collected. California generates approximately 360,000 tons of shredder residue each year, which ends up in landfills and can leach heavy metals into the surrounding environment. These heavy metals are harmful because they can contaminate surrounding groundwater tables and pollute community's drinking water, posing risks to human health. California's Department of Toxic Substance Control used life cycle assessment analyses to evaluate three alternatives to landfilling to minimize this leaching effect and identified one option—in this case, using shredder residue as fuel and mineral inputs into cement manufacturing—that emerged as a clear environmental winner (Boughton & Horvath, 2006). Specifically, recovery of shredder residue for cement manufacturing could save 1 million tons of coal from being produced each year in the United States, and the authors found that this method provided net environmental benefits across all impact categories evaluated in the study (Boughton & Horvath, 2006). LCA has been used in this way to determine the strategies that can be used to minimize a variety of environmental impacts, including greenhouse gas emissions related to global warming, stratospheric ozone depletion, acidification, photochemical smog, eutrophication, human toxicity, ecological toxicity, and resource depletion (US EPA, 2001).

The basic philosophy implicit in LCA is life cycle thinking, which considers all environmental impacts of a product or a material across its entire lifespan, instead of focusing on one specific impact or one stage of use. Today's policies are increasingly viewing environmental problems as interconnected challenges that affect air, land, or water quality across national borders. While today's environmental policies are becoming increasingly proactive, previous environmental policy-making was often corrective, or end-of-pipe, focusing on regulating selected environmental concerns (air pollution, water pollution, etc.) or individual life cycle stages (production, waste management, etc.) by applying a last minute technology fix to a technological problem. As Curran observes, "such narrow solutions run the risk of simply transferring impacts to another stage in the life cycle or resulting in little or no overall beneficial effect" (Curran,

1997, 39). Today's policy makers recognize the interconnectedness of environmental challenges, which grow increasingly complex as the world's economy integrates. LCA has the potential to aid in the institutionalization of a shift toward a more holistic and strategic approach to environmental management, because it integrates the impacts across time and space (Cowell *et al.*, 2002, 881). For example, when we decide to buy a car, we must consider not only the car's effect on local air quality but also its impact as a source of GHG emissions, the resources that were extracted to build the car, and how the car will impact the environment when it becomes a waste. LCA is designed to quantify the entirety of a car's impact and prevent environmental policies from merely shifting environmental burden from one type of impact to another or from one place to another. A report by the United States Environmental Protection Agency (US EPA) on LCA methodology uses a hypothetical situation to explain the advantage of LCA's comprehensiveness:

> An LCA allows a decision maker to study an entire product system, hence avoiding the sub-optimization that could result if only a single process was the focus of the study. For example, when selecting between two rival products, it may appear that Option 1 is better for the environment because it generates less solid waste than Option 2. However, after performing an LCA, it might be determined that the first option actually creates larger cradle-to-grave environmental impacts when measured across all the three media (air, water, land) (e.g., it may cause more chemical emissions during the manufacturing stage). Therefore, the second product (that produces solid waste) may be viewed as producing less cradle-to-grave environmental harm or impact than the first technology because of its lower chemical emissions (Curran, 2006, 3).

We see three main benefits of LCA versus other types of environmental indicators. LCA outputs (1) are able to quantify tradeoffs, (2) capture impacts irrespective of place and across different localities, and (3) measure impacts over the whole lifespan of a product or a service system. It is mainly for these three reasons that some experts feel LCA techniques will be instrumental in the 21st century phase of global environmental policy-making.

LCA has continually gained momentum as a potential policy instrument. Many governmental applications have been found for LCA studies, including product-oriented policies (eco-labeling and green purchasing); deposit-refund schemes, including waste management policies; subsidies and taxation; and finally, general process-oriented policies, like toxic substance management (Jensen *et al.*, 1997, 45). Ross and Evans establish the policy-relevance of an LCA, concluding that it "is capable of assessing the trade-off between several policy options implemented at various points in the life cycle and · · · reveal the policy mix that would lead to the most beneficial outcome for the environment" (Ross & Evans, 2002, 140).

On the basis of these perceived benefits, Cowell *et al.* find that especially in European countries, "today [LCA] is increasingly used to inform public decision making" (Cowell *et al.*, 2002, 886). They discuss three examples of LCA studies informing policy-making at the national and the continental level in Europe: the European Commission's Eco-Labeling Scheme set up in 1992, regulations of Polyvinyl chloride in packaging in the Netherlands and Sweden in the 1970s, and the waste management strategies in the UK from 2000. More recently, France has gone forward with a policy that requires all the products sold in the country to have an environmental product declaration, effective 2011 (Schenck, 2009, 2). It is expected that LCA studies will provide the informational basis for this particular eco-labeling scheme and that other EU member countries will follow suit (*ibid.*), expanding the role of LCA in European environmental policy discourse.

LCA has also been used in policy discussions and decisions in the United States, though it has less impact in the United States than it has in the European states. Several U.S. states and municipalities have successfully used LCA as a decision support tool (Schenck, 2009). Notable among these is the State of California's Office of Pollution Prevention and Technology Development, which has conducted a series of LCA studies, including the shredder residue assessment discussed above, to guide its hazardous waste management policies (California Department of Toxic Substance Control). Environmental groups such as Green Seal have moved forward in using LCA to frame their activities, and businesses commonly use similar assessments to lower the environmental and financial burden associated with their practices. But achievements of equal scale have not been made at the national level in the United States.

President Clinton's two terms in office witnessed a modest but definitive incorporation of life cycle assessment to help guide environmental policy at the national level. Under his leadership, the federal government changed its guidelines to allow for the use of LCA in guiding purchasing decisions. A report by the EPA summarizes the policy basis for the guidelines:

> In October 1993, President Clinton signed Executive Order 12873, "Federal Acquisition, Recycling, and Waste Prevention," which directs Executive Agencies to evaluate the environmental attributes of the $200 billion in products and services purchased by the Federal government each year. Executive Order 13101 entitled "Greening the Government through Waste Prevention, Recycling, and Federal Acquisition," signed September 14, 1998, further defines the Federal government's preference for "environmentally preferable" products and services (US EPA, 2000, 1).

Responding to the opportunity provided by the new federal mandates, the EPA developed the Framework for Responsible Environmental Decision-Making (FRED), which is "a practical methodology to guide environmentally preferable purchasing" (*ibid.*). Many states followed suit, mandating "life cycle analysis" considerations be used in purchasing decisions (Lowman, 1997, 4). Notably, the approach institutionalized in FRED "involves integrating price, technical performance and environmental information based on LCA into purchasing decisions" (1). Since the development of FRED, other government departments have begun using LCA in their policies. For example, the United States Forest Service uses LCA to make recommendations for design professionals to use reused as opposed to virgin wood products (Bergman, 2009). The choice of LCA to inform these recommendations was based at least partially on the growing acceptance of LCA internationally and domestically (*ibid.*).

Some argue that on the market-driven basis of this type of environmental policy, the use of LCA should be scaled up in U.S. environmental policy-making. Schenck argues, "the use of LCA as a policy instrument provides many opportunities for rational and cost-effective environmental decision-making and can provide substantial economic incentives to those organizations embracing environmental sustainability as a business strategy" (Schenck, 2009, 1). Looking at regulations under the Clean Air Act and the National Environmental Policy Act alongside other policy arenas, she concludes, "within the US, the broad use of LCA in regulations can decrease compliance costs and increase effectiveness...and its use in policy creates many ways for U.S. business to be globally competitive (5)." Such arguments demonstrate willingness for more widespread adoption of LCA studies and the principles in environmental policy debates in the United States.

However, adopting a more widespread use of LCA in U.S. environmental decision- making may not be easily achieved. Like most decision-making tools, the uptake of LCA in the political realm is affected by the structure of the American policy-making environment. In fact, some have argued that the very basis of LCA as an environmental indicator makes it incompatible with the environmental policy-making system in the United States. Bras-Klapwijk shows that the "development of the LCA methodology has been based, consciously or unconsciously, on a...normative view of sound public policy-making and analysis known as the rational theory" (1998, 335). The rational-comprehensive policy-making model claims that public policies are determined by the identification of targeted goals and an analysis of the optimal way to achieve those goals, comparing all available means. It is assumed that policy makers will be able to objectively evaluate among alternatives and pick the best one (Bras-Klapwijk, 1998). Such a model assumes a decision-making context where "The environmental planner is more of an apolitical, technical expert, striving to engineer a harmonious relationship between nature's and man's works to avoid irreversible damage and to secure the long-term viability of ecosystems" (Briassoulis, 1989). LCA was crafted with such a process in mind, and its success is therefore somewhat dependent on the presence of this type of policy-making. The nature of LCA methodology—its quantitative output, comprehensiveness across life cycle stages and cumulative impacts across geography—was tailored to guide policy makers who operate in a rational public policy setting. However, the reality is that the structure of environmental policy-making in the United States is incremental, not rational and comprehensive. Policy does not advance smoothly toward a clear optimal end state. This reality may cause occasional frustration, but the rules of environmental decision making necessitate that we consider this reality in designing political strategies for environmental issues.

In the rest of this chapter, I will discuss the nature of the U.S. policy-making environment as it relates to life cycle assessment. I will focus on the incremental nature of environmental policy-making and its deviation from the rational comprehensive model on which LCA was created. It is my contention that the U.S. environmental policy context must be taken into account when we seek to change the underlying methodologies of decision-making tools such as LCA. Otherwise, LCA runs the risk of becoming a tool that is increasingly analytically sound, yet remains unable to influence environmental policy in any meaningful way—or perhaps even makes policy outcomes worse (Bras-Klapwijk, 1998).

## The Policy-Making Context in the United States: Incrementalism

The United States policy-making more often resembles a meandering series of policy steps than a linear trajectory toward a specific policy goal. It can be said that the political setting in which modern American environmental policy-making takes place is defined primarily by incrementalism (Brulle, 2010, 11; Repetto, 2006). Environmental policy progress is not made steadily, but typically happens in fits and starts. The reality that the policy advances through "piecemeal gradualism" contrasts with the idealized vision of the policy-making process that characterizes the rational model of environmental politics (Lindblom, 1979). In this model, the outcome of environmental policy should be the "selection of the best solution that meets objective scientific criteria" (Briassoulis, 1989, 384). Life cycle assessment fits nicely into the rational decision model because outputs identify a clear environmental winner between a set of products or services (Bras-Klapwijk, 1998). The expectation is that the results from an LCA will be taken as a factual statement of environmental preferability and then used to design an optimal environmental policy. However, in American politics, policy decisions require crosscutting consensus across ideological lines, and this demand for consensus tends to guide policy makers cautiously away from environmental damages, rather than toward an agreed-upon environmental goal, complicating the seemingly simple process of choosing the least environmentally harmful product option (Braybrooke & Lindbolm, 1963, 71). Given the disjointed and incremental advance of U.S. environmental policy, it is unsurprising that LCA as presently practiced has not achieved serious political relevance in the United States.

The structure of the U.S. government in many ways conditions the incremental policy-making approach. Lindblom observes a "structure of veto powers," both explicit and implicit, which guide society toward incremental political moves (1979, 520). The process of reconciliation between the House and the Senate guides policies toward consensus rather than fundamental and structural policy reforms (Selin & VanDeveer, 2011, 15). Separation of powers—the fact that multiple branches of government are open to challenge or obstruct environmental policies—can also contribute incrementalism (Repetto, 2006, 3). Plus, this type of policy-making suits the needs of politicians, as it allows them to minimize risk by taking small steps in reaction to an environmental issue, instead of taking deliberate, large leaps that might marry an individual to a particular issue. Politicians' own time horizons are based on the electoral cycle and are much shorter than the time horizons over which long-term environmental challenges play out, which often causes policies to focus on short-term, instead of long-term benefits. As Hammond *et al.* put it, "Policy makers have high career stakes in the success or failure of policies for which they are responsible" (Hammond *et al.*, 1983, 20), and this creates an inherent risk-aversion that is embedded into the political process.

However, incrementalism does not in itself preclude the passing of effective environmental policies. According to Lindblom, "incrementalism in politics is not, in principle, slow moving... A fast-moving sequence of small changes can more speedily accomplish a drastic alteration of the status quo than can an only infrequent major policy change" (1979, 520). His line of reasoning stems from the very basis of incremental politics: because it by definition does not stray from the existing consensus, incrementalism as a policy feature minimizes the polarization and values-based hostility that more drastic reforms may stir up. A classic example of the latter approach is the current state of U.S. climate change politics, which inflames much partisan rhetoric while accomplishing very little on the ground. Therefore, though smaller in scope, incremental policy outcomes can come about at a relatively consistent pace, which has ramifications for our current slate of pressing environmental issues. Applying this notion to climate change, Levin *et al.* (2007) devise a strategy that they term "progressive incrementalism." (2007, 11).

They seek to upset the current stalemate on climate change policy through the implementation of carefully selected, yet incremental policies that would counter current practices and "lock-in" more sustainable ones. Their strategy hinges on the notion of path-dependency in policy-making: "The goal of our approach is to identify those policies that once enacted are both difficult to reverse, and in the most promising cases, become more entrenched over time" (13). Therefore, it is false to assume that incrementalism closes off the opportunity to address environmental issues in a meaningful way. The challenge for the LCA community is to accept the reality of incremental policy-making while crafting new environmental information that fits into the incremental pattern by insulating itself from the perils of partisan politics.

## Non-Incremental Leaps in Policy

As we saw in the flurry of environmental legislation passed during a short period of the 1970s, great environmental policy leaps and policy innovation are possible. Braybrooke and Lindblom refer to "grand opportunities," which can often jolt the policy-making process out of the pattern of incrementalism. For example, Charles O. Jones, writing in 1974, observed that the 1970 Clean Air Act was an example of a non-incremental leap over previous air quality legislation (438–464). In his view, the increase in public concern and a change in the public's attitudes regarding air quality created a "grand opportunity" for change and for successful problem solving. What had been a poorly understood and tentatively addressed concern was transformed by media attention and popular awareness into a solvable policy problem. While we saw this with the 1970 clean air legislation, we did not see it in 2010 when the U.S. Senate considered enacting climate legislation.

Certain environmental issues are able to gain enough traction to leapfrog the incrementalist status quo. Repetto writes about the existence of "positive feedback loops" hidden within the structure of environmental policy-making that can interact to shift the system away from incrementalism, if only temporarily (2006, 11). The presence of these feedback loops gives the policy-making system the capacity to allow some environmental policy issues to advance non-incrementally. Holt and Barkmeyer stress the importance of media in this process, particularly its "vital dual role in agenda setting and transmitting information" (2010). If people perceive that a problem seriously threatens their well-being or the well-being of their families, then an environmental issue can become a major factor in national politics. But it is not enough for scientists or other experts to "know" that the problem looms—the public must witness and feel its daily effects, which often requires media attention. Others point to positive feedback loops within the political process itself. One important theme is the role of "political entrepreneurship" in bridging the gap between the public and the policy-making apparatus and consolidating these deviations from the incremental standard (Lyons, 1999, 290; Repetto, 2006, 11). Referring back to the Clean Air Act and other landmark environmental laws of the 1970s, Lyons argues that, even as the visibility of issues was rising, the role of Senator Edmund Muskie's policy entrepreneurship provided critical leadership in transferring the momentum into legislation (1999, 290). Political entrepreneurship consists of the active mobilization of public sentiment on the part of enterprising political figures, despite the associated political risks, and helps increase the political salience of environmental issues that concern the public. Leadership at these critical junctures can become institutionalized given the prominence of "bandwagon effects" in American politics (Repetto, 2006, 11). Thus, the political process retains the capacity to dynamically address certain environmental issues.

Brulle identifies four main drivers of environmental politics, creating a framework with which to consider this dynamic issue: the political opportunity structure, the structure of the environmental movement, cultural dynamics in relation to environmental issues, and mobilization in response to environmental conditions (2010, 9). While incrementalism may generally inhibit major changes in legislation, changes along any of the four dimensions, if sufficiently powerful, can restructure the nature of environmental politics in relation to that issue. Brulle provides a particularly potent example, given the 2011 earthquake and tsunami in Japan: "early nuclear accidents were virtually unnoted by the mass media and policy makers in the 1950s (e.g., the Fermi near disaster), creating no public response, but Three Mile Island and Chernobyl stirred considerable protest and mobilization in the 1980s. This was due to a reframing of nuclear incidents as potentially catastrophic in nature" (2010, 12). Despite the presence of incrementalism as a structural attribute of the environmental policy-making context, the relationship between America's politics and the environment is far from static and often changes on the basis of public interpretation of environmental events.

The notion of "punctuated equilibrium" has emerged as an influential theoretical model that integrates the incremental and non-incremental features of environmental policy-making in the United States into one formula. Borrowing from evolutionary biology, Baumgartner and Jones (1991) adapted the concept and applied it to explain the pattern of general stability coupled with rapid change evident in many policy areas. Despite the infrequency of non-incremental leaps, the authors show, "Often, the grand lines of policy may be settled for decades during such critical periods of mobilization" (1991, 1044). In this model, the discontinuous breaks from incrementalism discussed above serve as "focusing events" or as Birkland (cited in Brulle, 2010) explains, external stresses that "change the salience of issues and sometimes replace indicator-based analyses with much more emotionally charged examples of policy failure and the need for reform" (Brulle, 2010, 12). Such events can increase the political visibility of an issue and funnel the attention of the public and policymakers toward environmental reform, disrupting the incrementalist equilibrium. Thus, the punctuated equilibrium model focuses on the "interaction of beliefs and values concerning a particular policy...with the existing set of political institutions" (Baumgartner & Jones, 1991, 1044). This dynamic model shows that both incremental progress and non-incremental leaps are fundamental components of American environmental politics. Understanding the overall environmental policy-making process, including both incremental and non-incremental events, will help life cycle–assessment practitioners create LCA outputs that may be more useful in translating environmental data into policy. An interesting feature to keep in mind is the one that makes intuitive sense but is often ignored by those caught in the environmental policy debate: people react much more strongly to events that are emotional and personal than ones that are numerical and empirical. While this flies in the face of rational public policy as well as those trained in the scientific method, it does allow for creative use of environmental information, especially if we consider the scale at which environmental issues tend to be felt the strongest.

## Place-Based Environment Politics: The Importance of Scale

While many of the most challenging environmental problems are global or national in scale, the U.S. policy-making system is typically not conducive to widespread problems. This is because the environmental policy-making system in the United States is place-based in nature. The United States' political institutions respond best when an issue is extremely visible on a local level. Take, for example, the overwhelming media and policy responses to the 2010 oil spill in the Gulf of Mexico. The magnitude of the response to the oil spill grew out of the fact that its impact had a direct effect on the lives and livelihoods of local citizens. Because of this response, plans to expand offshore drilling as part of the effort to boost domestic energy production were temporarily put on hold. On the other end of the spectrum of environmental impacts, we have climate change or marine debris, two impacts that affect the entire world's population but are less obvious on a local level and therefore don't make it onto the agenda. While international topics such as these are discussed at length and with increasing frequency in the online media, when it comes to real action at the policy level, local issues often take precedence. Thus, we witness serious resistance to the adoption of climate change legislation in the United States, despite the general public consensus in favor of environmental protection. It is possible that recognition of climate change as a cause of serious local disasters, such as Hurricane Katrina or the increasing number of tornados in the South and Midwest, could generate momentum behind U.S. national climate change legislation. But this attribution is fraught with scientific uncertainty and the politics that stem from it, a topic I will discuss at length in the next section.

It is difficult to overemphasize the power of the place-based nature of American environmental politics. Fear of environmental damage has resulted in a "not in my backyard" (NIMBY) syndrome that "often produces greater total environmental impacts in order to avoid lesser effects on a more powerful or better organized local constituency." Take, for example, the issue of nuclear waste. After the federal government has spent billions of dollars to develop and complete a nuclear waste repository in Yucca Mountain, Nevada, the Nevada delegation to the U.S. Congress, vetoed its operation. As I wrote for *The Huffington Post*:

> The "Not-in-my Backyard" (NIMBY) syndrome is a central element of land use politics in communities throughout the United States. While it is true that the definition of a noxious facility varies from place to place, no one doubts the ability of an American locality to veto a land use that they do not like. In New York City we have an extreme version of NIMBY where we even have trouble siting big box retailers. Most places are happy to allow Wal-Mart, but even before last week, few communities were interested in hosting a nuclear power plant The strength of anti-nuclear power politics should not be underestimated. In New York, people on Long Island are still paying off $3.3 billion in debt for a nuclear power plant called Shoreham that, like the Yucca Mountain repository, was completed but never opened (Cohen, 2011).

There are several explanations for the place-based bias in environmental policy. Lyons expects more widespread issues like climate change to remain in a state of political stalemate, while localized environmental issues like wastewater treatment more readily achieve political agenda status and gain traction in terms of policy outcomes. According to him, this is because of the structure of political benefits that politicians gain by pursuing certain types of environmental policies. Lyons explains, "the U.S. political system offers to politicians abundant incentive to provide tangible and specific policy benefits, yet relatively little incentive to provide benefits that are diffuse or intangible." (1999, 275). By the very nature of the functioning of the U.S. government, issues that are tangible and local are privileged over widespread ones.

Providing a different perspective, Selin and VanDeveer look at this result more structurally, and with a global perspective, highlighting the way multilevel environmental governance, such as that in the United States, which includes layers of federal, state, and municipal governments, gives political space for local actors to take the lead on environmental issues. Multilevel governance refers to the interaction of "multiple public and private sector actors operating across horizontal and vertical levels of social organization and jurisdictional authority" (2011, 6). Specific to the multilevel nature of climate change governance, the authors identify federal, state, and municipal policymakers, private sector leaders, and civil society representatives as the key players. Within the system of multilevel governance, a country's domestic politics can greatly influence how outcomes are determined (Andonova & Mitchell, 2010, 260). In the US, it can be argued that "environmental federalism," or the division of environmental policy-making roles between municipalities, states, and the federal government, is a primary characteristic of multilevel governance (Selin & VanDeveer, 2011, 12). Such an apportionment of roles has arisen primarily due to policy inaction at the national level (Selin & VanDeveer, 2011, 12) and can be seen as part of an active process of "rescaling" between various tiers of the multilevel governance structure (Andonova & Mitchell, 2010). Rescaling can be defined as "a shift in the locus, agency, and scope of global environmental politics and governance across scales," and often demonstrates that there exists a level of competition between various actors at different levels of society for increased power and control (Andonova & Mitchell, 2010, 257). As Selin and VanDeveer write, in American environmental politics, rescaling is linked primarily to competition between the various levels of government: "Since its inception, the United States has been home to contentious debates and relations between the federal government and state governments competing for legal authority and resources" (2011, 12). This theoretical approach shows that given the scaling down of environmental governance in the United States, it is only natural that local issues would be responded to more quickly than national or global ones.

The relatively modest political relevance of LCA becomes easier to explain when considering that the place-based emphasis of U.S. policy-making forms an ingrained and integral aspect of environmental policy-making. As I described earlier, LCA as a scientific input was designed to provide a holistic and comprehensive indicator that could be used in policy debates. Particularly, LCA is meant to capture environmental effects that span geographic and political boundaries. As we have seen, however, the political system is likely to produce policy solutions to local environmental problems, but structural constraints preclude the same type of consensus around global and widespread problems. Given this political context, even if practitioners perfect the ability of LCA to capture trans-geographic environmental impacts, it is not expected that LCA will become much more effective at influencing the policy debate. Instead, it might be more effective for LCA practitioners to tailor their studies so as to enable them to specify environmental damages at specific geographic scales. This disaggregation would allow the quantified environmental damage to enter a place-based dialogue on environmental policy, possibly moving an issue onto the non-incremental, transformative track.

## The Role of Uncertainty in U.S. Environmental Policy-Making

The scale of environment problems and the corresponding local, state-wide, or regional politics only partially explain the difficulties LCA practitioners have encountered in using their outputs to influence policy outcomes. The incremental and localized nature of environmental policy-making in the United States comprises one set of constraints on the use of scientific inputs in policy-making discussions. We must also consider the nature of scientific information as another important obstacle that hinders the easy uptake of indicators such as life cycle assessments. By their very nature, environmental processes are complex and intertwined. In the real world, these complexities must be simplified in order to clarify the ways human activities influence natural systems. A certain degree of uncertainty is a necessary element of any quantitative environmental indicator. Science has evolved hand-in-hand with the presence of these uncertainties, which explains why, for instance, climate models publish their results in terms of probabilities and likelihoods—a given climate prediction is deemed to be *very likely*, or *somewhat likely*, but never certain. While scientific understanding accommodates uncertainties, politics and policy-making are less ideal venues to incorporate uncertainties into their processes. Summarizing the problems this poses for politics, Hammond *et al.* write:

> Scientific statements, particularly projections of the likely future impacts of alternative policy options, are ordinarily probabilistic, complete (one hopes) with clearly stated confidence limits. Policy makers' requests for policy-relevant scientific information can seldom, if ever, receive precise, unequivocal answers. Thus, scientific information does not mesh well with policy makers' needs to make singular, discrete choices. Policy makers frequently find themselves in situations analogous to that faced by all of us at times when we must decide whether or not to carry an umbrella on a day when the weather bureau has forecasted a 20% chance for rain. (1983, 289)

Uncertainty is an intrinsic ingredient in our understanding of any scientific problem. The problem of uncertainty is more pronounced for environmental issues due to the layers of complexities present within ecological systems themselves as well as in how human and ecological systems interact. For these types of deeply interconnected systems, typical tools like the scientific method are impossible to employ because they rely on controlled variables, and there are hundreds of uncontrollable and unobservable variables present in any environmental system. Since we cannot understand what an undisturbed system looks like, quantitative data regarding changes in disturbed systems typically is inaccurate and fails to capture possible variables (Leinfellner, 1990). Furthermore, the complexity of environmental problems means simplification is a necessary component of any attempt to comprehend environmental processes—we move from the solidity of controlled experiments to the ambiguity of scientific and mathematical models. Because environmental problems are not readily reducible, choices must be made in simplifying the systems, even when we use mathematical models, and these choices can easily influence the subsequent result. As Arrow *et al.* explain, "We are relatively ignorant concerning relationships in ecosystems and are likely to underestimate the list of services they provide" (2000, 1402). While some environmental problems are scientifically simple, many environmental processes are composed of hundreds of reactions and relationships. Overlooking even one key relationship could dramatically alter the veracity of scientific outputs. Any quantitative measurement meant to convey even a portion of the total environmental impact of a man-made good or service—be it a soda can, a particular emission from a factory, or an entire transportation network—must first navigate a series of potentially distorting simplifications and assumptions.

Environmental science—and more specifically, the indicators used to demonstrate its findings—is in this way precluded from making statements with absolute certainty. Scientific and quantitative tools and techniques are constantly evolving and improving but will never deliver results completely devoid of ambiguity. It is typical to find that when harm cannot be proven, a lack of certainty often inhibits environmental protection policy. According to former United States EPA Administrator William D. Ruckelshaus, "...EPA's laws often assume, indeed demand, a certainty of protection greater than science can provide with the current state of knowledge" (1983, 1026). This highlights the importance of uncertainties attached to the proposed policies themselves. Because environmental systems are so complex, we cannot be 100% certain that a given policy will produce the intended effect without any serious side effects. This lack of certainty gives ammunition to those who are ideologically or politically opposed to environmental regulation. As our understanding of the natural world improves and as our mathematical and computational capacities increase our ability to model numerous interconnected processes and feedback loops, the challenges of uncertainty can be lessened. Improvements along these fronts would in turn influence how information is used to address environmental issues (Layzer, 2002, 230). A brief look at climate change policy in the United States under President George W. Bush makes the point. During the early years of his administration, his top advisers perceived a lack of scientific certainty about the scientific basis of global warming. As scientific consensus emerged and uncertainty was reduced, the government had more incentive to address global warming (Cohen, 2006, 30). Thus, refining the methodological aspects of LCA to make the findings more scientifically robust and less uncertain could facilitate the instrumental use of LCA in the U.S. environmental policy-making debates.

But as long as uncertainty remains, the scientific problem of environmental issues will always be at least partially subjective. Subjectivity goes hand-in-hand with simplification in the production of environmental information. Importantly, this subjectivity exists even if individual scientists adhere stridently by scientific and ethical principles. The development of life cycle assessment, both methodologically and as a political input, confirms the way subjectivity and value systems can alter the results from supposedly unbiased environmental indicators.

Subjectivity is built into the very premise of life cycle assessment. Recall the four phases of the standard LCA study: goal and scope definition, life cycle inventory analysis, life cycle impact assessment, and interpretation. Finnveden (2009) points to three sources of uncertainty that reside within these phases:

- *data*, e.g., electricity use of a heating boiler, $CO_2$ emissions from a coal-fired power plant, and GWP of dinitrogenoxide;
- *choices*, e.g., system boundaries, allocation principles, and time horizon in Impact Assessment;

- *relations*, e.g., the linear dependence of traveled distance on fuel input, the linear dependence of acidification on $SO_2$ emissions, and the discounting formula used for long-term impacts (2009, 14).

Uncertainty within these three sources can create difficulties in the transmission of environmental information into the policy-making world. For example, Cowell *et al.* make the case that "LCA results can be complicated and may require specialist knowledge for their interpretation. It can be difficult, therefore, for nonexperts to debate the validity and relevance of LCA studies in a decision-making situation" (2002, 888). In this way, uncertainty and complexity impair the ability of politicians to understand scientific outputs, given the technicality of the measurement.

In addition, there exist certain unique types of uncertainty in LCA, which necessitate subjective decisions on the part of practitioners, such as the decision whether to include water use in an inventory analysis (Finnveden, 2000, 231). For issues such as these, the importance of choices in LCA analyses increases the prominence of subjectivity on the part of practitioners. Finnveden (2000) surveyed the inventory analysis phase of various LCA studies and found that they typically "do not cover all relevant environmental aspects." As a consequence, "No conclusions can be drawn concerning the overall preference from an environmental impact perspective of one choice over another, simply because all environmental aspects have not been included" (Finnveden, 2000, 231). In this way, the selection of environmental impacts to measure constitutes a type of subjectivity in LCA analyses, limiting their capacity to provide objective policy advice.

The use of different value systems in environmental policy-making is inherently problematic. And one of the more challenging aspects of LCA is its reliance on value interpretation, which complicates the utilization of LCA studies in policy debates. Cowell *et al.*, noting the increasing awareness of value judgments within LCA methodology, highlight three areas of LCA applications that necessitate value judgments on the part of practitioners. Most basically, philosophical questions emerge. While "LCA is based on the premise that tradeoffs can be made between different environmental impacts," some argue that such tradeoffs are in principle not applicable for certain irreversible types of environmental damage (Cowell *et al.*, 2002, 887). Looking at the interpretation phase of LCA, the authors find, in line with Finnveden (2000), that this process requires the adoption of value judgment about the relative desirability of different types of environmental impacts (Cowell *et al.*, 2002, 888). Lastly, looking at the nature of LCA results, they observe, "LCA is not concerned with inherently qualitative, subjective aspects" that some may consider valuable aspects of the environment (Cowell *et al.*, 2002, 888). For example, returning to the LCA impact categories created and formalized by the EPA: global warming, stratospheric ozone depletion, acidification, photochemical smog, eutrophication, human toxicity, ecological toxicity, and resource depletion, the exclusion of other potential categories, such as animal welfare, was an inherent choice that is in part a value judgment.

The subjective elements inherent in life cycle assessment could comprise a venue for contestation of LCA-based environmental information in the political realm. In fact, Bras-Klapwijk (1998) shows how the inherent subjectivity of LCA results, when coupled with an expectation of quantitative objectivity, can easily lead to polarization in political debates that can actually result in worse policy outcomes. Even when uncertainties are explicitly presented, experience of LCAs in the political realm:

> ...shows that political actors suppressed or denied qualifications when the results were in line with their view. On the other hand, actors tended to emphasize these uncertainties when they did not agree with the result. Qualifications in LCAs often led to polarization about the quality of the study (Bras-Klapwijk, 1998, 340–341).

Arguments can always be leveled against the quantitative output of an LCA because of the inherent uncertainty of the process. While "the result from an LCA is a single observation statement," subject to different outcomes under different parameters, "a statement that one product is environmentally preferable to another one is a universal statement...Anybody who wants to challenge the results from an LCA can always ask for a new situation with slightly different properties that were not included in the original calculations" (Finnveden 2000, 234).

In this way, the fact that by trying to reduce incredibly complex systems into comprehensible pieces inevitably creates a level of uncertainty means that by their nature, life cycle assessments will necessarily be limited and constrained. Given these potential pitfalls and the possibility of political manipulation, it is necessary to consider various ways the U.S. policy-making system copes with uncertainty in crafting environmental policies, so as to move LCA outputs to better fit into the framework of environmental policy-making we observe in the world.

## Case Study: Using U.S. Climate Change Politics to Explain the Limits of LCA

The case of climate change politics in the United States provides insight into the functioning of the overall system of environmental politics in the United States as I have laid out above, and it also highlights the ways this political structure inhibits the use of life cycle assessment in the surrounding policy debate. Ultimately, the dynamic of climate change politics disempowers LCA as currently conceived, by minimizing the effectiveness of science-based, rational indicators in policy-making. This is because the widespread uptake of environmental indicators hinges on values and the prevalent discourse with respect to the environment, which at present are highly polarized. Climate change deniers and opponents with political agendas have framed climate policy as zero-sum: claiming that effective climate policy can only be pursued at the expense of domestic economic prosperity (Cohen, 2006, 109). As a result, we find the debate remains stuck in a contentious stalemate, even as polls published in May 2011 by Yale and George Mason Universities found that 64% of Americans now believe that the planet is warming—an increase from similar polls conducted last year (Leiserowitz *et al.*, 2011, 2). But politicization has clearly affected the issue, as more Americans believe that disagreement rather than consensus among scientists characterizes the science that provides evidence of climate change (*ibid.*, 3). This kind of public sentiment increasingly mirrors the climate views of leading politicians, who project this perceived disagreement into the debate around climate change politics. Such views are perplexing, considering that scientific consensus is only growing, as evidenced by the publication in the journal *Science* of a letter signed by 255 of the world's most distinguished scientists defending the integrity of climate science (Gleick *et al.*, 2010). This situation of politicization and polarization explains the lack of transformative politics on this issue—confirming Brulle's contention that in the presence of severe polarization between the pro- and anti-change communities, potential focusing events such as new or improved scientific information will still fail to dislodge environmental politics from the incremental equilibrium (2010, 12).

My main argument is that climate change predominates not as a *scientific* issue, but as a *political* issue. The scientific community has commendably enhanced the quality of its work in projecting climate impacts of human behaviors. But this development has mostly occurred in isolation from the politics of climate change, which have only grown more divisive and should be considered separately from the science side of the issue. It is the political element of the issue that currently confounds scientists and environmental advocates alike. When thinking about environmental problems from a political perspective, it becomes clear that the political element explains why the numerous LCA studies on climate impacts of various aspects of our economy have not influenced the debate in any meaningful way.

It is not hard to imagine what transformative politics on climate change would look like. All countries, especially the most developed ones, would cut their greenhouse gas emissions substantially. This requires some combination of taxes or disincentives on the use of fossil fuels, subsidies to encourage the expansion of renewable energy industries, and lifestyle changes. Such changes might require a departure from the status quo, that is, from incrementalism. Many potential events could have shifted the United States out of its deadlock on climate policy, including international mobilization around the Kyoto Protocol or the 2009 Copenhagen conference, disasters like Hurricane Katrina or the recent tornadoes in the American Midwest and South, or new information and science as disseminated by the Intergovernmental Panel on Climate Change. But they haven't. Instead, we see the traditional incremental progress on this issue, which defines much of environmental politics; and as is common with incremental politics, the result is a meandering type of trajectory, defined by contentious politics between competing interest groups.

Many of the factors I described in the section on incrementalism have combined to give climate politics the disjointed "fits and starts" trajectory common to many environmental problems. These include electoral cycles and party politics, interest group lobbying, separation of powers, ideology, and heightened polarization. Consider the contours of U.S. climate change politics over the last 15 years:

- In 1997, the Kyoto Protocol was enacted at the third meeting on the Framework Convention on Climate Change. President Clinton agreed that the United States would participate in the agreement, which bound countries to emissions reduction targets averaging 5.2 percent below 1990 levels (United Nations Framework Convention on Climate Change, 2002).
- However, in anticipating of the President's accession, the Senate passed the Byrd–Hagel Amendment by a vote of 95 to 0 requesting that the United States not enter into any treaty requiring reductions that might damage the economy or hold developing nations to different commitment standard than developed nations. The Kyoto Protocol never reached the Senate.
- As of 2000, despite this inaction, the EPA accepted the fact of anthropogenic climate change as part of its official policy (US EPA, 2000).
- In 2001, President Bush formally pulled the United States out of the Kyoto Protocol.
- By 2002, he adopted a climate policy related to reducing the "greenhouse gas intensity," or the ratio of greenhouse gas emissions to economic output, of the U.S. economy. In this way, policy on climate change was integrated into economic policy and was deemed secondary to economic growth, reflecting the power of pro-business lobbying in his administration. No specific reduction targets were set. But this policy indicated the relevance of climate change to national policy debates (Cohen, 2006, 105).
- In 2003, John McCain and Joseph Lieberman introduced the McCain–Lieberman Stewardship Act, which required domestic, mandatory, and economy-wide emission reductions, but the bill was defeated by a 55 to 43 vote later that year (Pizer *et al.*, 2003, 1). Again, the lobbying of energy industry and inflamed rhetoric on both sides characterized the politics of this bill (Cohen, 2006, 114).
- After a period of Congressional inaction, the election of President Obama in 2008 seemed to signal a change in climate politics. In 2009, the House of Representatives passed the American Clean Energy and Security Act, which promoted a clean energy economy and a cap-and-trade system. A year

later, the American Power Act was proposed in the Senate but failed when Democrats removed all meaningful carbon emissions legislation in order to appease Republicans and pass the bill (Hulse & Herszenhorn, 2010, 1). The passage of sweeping climate change legislation remains one of the main unfulfilled promises from Preisdent Obama's campaign.

- However, advances have been made in terms of climate change policy outside the legislative branch of government. In 2007, the U.S. Supreme Court voted to require the EPA to regulate greenhouse gas emissions as air pollutants under the Clean Air Act. The EPA is now set to regulate greenhouse gases from factories and power plants in a policy that has been under development over the past several years. Such measures are vehemently opposed by Republicans in the House and Senate, who have convened hearings intended to call into question the foundations of climate science (Revkin, 2011).
- In 2009, President Obama used his regulatory authority to "significantly increase" fuel efficiency standards for automobiles through 2016, with assent from the automotive industry. He is preparing to ramp up these standards through 2025 (Harwood, 2011).
- In his 2011 State of the Union address, President Obama set down his climate policy outside the framework of legislative politics. Though this policy falls short of a climate bill that supporters have been advocating, he did set many ambitious climate-related goals during the speech, including the manufacture and sale of 1 million electric cars in the United States by 2015 and the target of producing 80% of the nation's energy from clean energy sources by 2035. This shift in focus represents an effort to wed environmental policy to economic policy and an overall strategy of 21st century global competitiveness (Cohen, 2011).

Clearly, climate change policy has evolved and progressed since the Clinton era, but just as clearly, it has yet to make the leap of progress at the level of the federal government, and many experts feel we need to avert the worst effects of a warming planet. The main explanation for this is that climate policy does not mesh well with the place-based nature of American environmental politics. Those environmental issues with grassroots support, ones that generate clear, sustained, and visible impacts felt at the local level, are more likely to achieve political salience and generate momentum for policy solutions. However, climate change is remote and distant, and its impacts are imprecise and based on mathematical models and projections. The major venue of climate politics is between scientists, corporations and politicians, and lobbying elites seasoned in Beltway politics. Furthermore, even though it is widely noted that comprehensive climate legislation would very likely generate net societal benefits outweighing the costs and that the lack of policy is harming the U.S. competitiveness internationally, the shift would be costly for select industries. The resulting job losses would be felt very strongly at the local level in the areas where they are sustained, and this issue, especially during a recession, is politically salient. This increases the political risks to individual politicians for pursuing a climate policy. Another interesting consequence of place-based politics on climate change policy is that nuclear energy, formerly a clean energy option favored by both President Obama and many Republicans, became a political non-starter after the 2011 nuclear crisis in Japan, reducing the options for a bipartisan consensus on renewable energy. As discussed above, the fear of nuclear catastrophe is acute at local levels, and the siting of new plants will be almost impossible given local land use politics. Local issues like jobs and land use win out over widespread problems like climate change.

The uncertainty inherent in science also serves to open up climate change politics to competing influences, controversy, and debate. Climate change projections and attribution, like most science, have at least some degree of uncertainty. This uncertainty can then be manipulated by those opposed to climate change policy, disputing the science while emphasizing the tradeoff between action on climate change and economic prosperity. In this way, due to the very nature of scientific knowledge, issues at the interface of politics and science can always be challenged. The basic psychological impulse to suppress or downplay information at odds with one's own political views, and to seek out information that confirms those views, is most certainly at work in climate change politics.

The politicization of climate science, which grows out of the basic uncertainty of scientific information and the complexity of the climate system, has contributed to a polarization around the climate change issue. Lacking clear localized effects, the place-based preference of our political system renders it difficult for climate change to make the non-incremental leaps we have observed in some other environmental policy issues. However, as demonstrated by the history of U.S. climate change legislation, the United States has made strides on climate change over the past two decades, and it seems appropriate to characterize its evolution as a variant of "progressive incrementalism."

Climate change politics in the United States is one of the most contentious and ideological issues on the political agenda, with powerful interest groups mobilized on either side of the issue. A main aspect of the problem is the tradeoff between economic and environmental well-being, which is rooted in values. Rein and Schön (1993) argue that in this type of policy problem, the issue of framing becomes even more important than evidence (1993, 145). In policy debates, a frame is "a perspective from which an amorphous ill-defined, problematic situation can be made sense of and acted on," and usually integrates values, theories, and interests of the relevant political actors in addition to facts (Rein & Schön, 1993, 145). The way environmental problems like climate change are interpreted, beyond the fact of the problem itself, depends very much on an individual's setting and worldviews. This interpretation guides the decisions on what types of policy solutions (if any) should be adopted in response. Going forward, LCA studies would benefit from a more precise political understanding of this type of behavior, since it dictates the way LCAs are received by policymakers, politicians, and the public. A more nuanced appreciation of the interaction between science and politics would steer LCA applications away from the pitfalls of polarization, and toward a new consensus, however so slight that consensus might be. Recall Levin *et al.*'s pursuit of progressive incrementalism. Even if the first step toward climate policy is small, if arises out of genuine consensus, it can "lock in" more influential action down the road. That will require LCA practitioners to produce outputs in a format that anticipates and adapts to the criticisms of current detractors.

The climate science community is now starting to understand their role in this process of political action on climate change. A recent editorial in *Nature* (2010) acknowledges that the science community is in a "street fight" over climate science, and that, in light of the politicization of climate science, "scientists must not be so naïve as to assume that the data speak for themselves." (2010, 141). The editorial goes on to propose an enhanced relationship with the media, including media training and relationships with public relations firms (*ibid.*). This is in line with similar efforts within the science community, criticizing the "lack of scientific messaging" while calling for the development and implementation of "a serious communication strategy" so as "to effectively engage the public on this serious issue" (Romm, 2009, 23). Both the politics and the science around climate change seem grounded in political realities and poised to drive forward despite the current state of polarization of the issue.

## Conclusion

In this chapter, I have examined the way that life cycle assessment plays into the environmental policy-making system in the United States. As we have seen, the incremental nature of policy-making, the place-based structure of the U.S. political system, and the role of scientific uncertainty often preclude the successful use of LCA in real-world decision making. I then showed how climate change politics fit into this template of environmental policy-making in the United States. The United States policy-making system does not easily respond to a problem such as climate change in a neat, linear, and rational format. Climate change as an environmental issue is widespread, diffuse, and indistinct, occurring at a future time with unknown magnitude. However, the political context in which environmental decisions are ultimately made favors issues that have clear impacts at distinct local scales. Furthermore, the uncertainties in terms of the effect of and the ideal response to climate change have created the conditions for highly contentious, partisan, and often values-based debates, and we see the median points of either camp drifting further and further apart. At times, this no doubt frustrates those who seek strong, comprehensive, and forward-seeking action. But it has been my contention throughout this chapter that the best strategy, for political scientists who are concerned about the future health of our planet as well as environmental scientists and practitioners, is to accept the constraints posed by the political structure and to devise new techniques that can bring about the intended outcome while operating within the given political context. That is what I suggested to improve LCA, with one major forward step being an explicit incorporation of economic considerations.

Such a strategy may at first seem like a compromise, but in fact it maximizes the chances of ultimate success while focusing the efforts of the environmental community toward more productive ends.

I will conclude the chapter by discussing very briefly how a new and improved type of LCA, built on principles that accept the power and meet the requirements of prevailing political systems, could potentially begin to shift the environmental discussion in this country toward more fruitful ends. Practitioners of life cycle assessment must follow the changes within the political world and the climate science community and find ways to contribute to a reduced level of polarization in order to play a bigger role in climate change policy and beyond. So far, the use of scientific information has only further polarized the issue of climate change by politicizing the production and dissemination of science. But I have argued that progress on this issue requires a shared vision of the problem by all political actors. Thus, to participate more fully in this debate, LCA practitioners should keep this goal in mind and proceed to improve the methodology so as to enlarge the "formula" or shared definition. In my mind, the only way to achieve this is to address the critics and skeptics head-on. Perhaps LCA would be more successful if it explicitly incorporated economic assessment and demonstrated that environment-friendly policies would also be economy-friendly in the long run. I agree with Gregory A. Norris's critique of LCA, in which he argues that the "traditional separation of life cycle environmental assessment from economic analysis has limited the influence and relevance of LCA for decision making, and left uncharacterized the important relationships and trade-offs between the economic and life cycle environmental performance of alternative product design decision scenarios" (Norris, 2001, 118). By formalizing and institutionalizing economic considerations in LCA studies, LCA practitioners would undercut those who assail environmental policy on the grounds of it being too costly. In this way, advocates and policymakers seeking to implement certain environmental policies on the grounds that they would lessen greenhouse gas emissions could use LCA with economic analysis to justify the policy on economic grounds as well as environmental grounds.

The case of climate change demonstrates the difficulties of enacting climate legislation in the U.S. policy-making setting, but it also shows how LCA could be adapted to help reduce uncertainty and contribute to political action on climate change. In this chapter, we have seen that sometimes it is politics, not science, which impairs the current state of climate change politics. Ultimately, LCA could contribute to a process of reconciliation and consensus building between political opponents and in doing so would become very instrumental in U.S. climate policy.

## References

Andonova L., & Mitchell, R. B. (2010). The rescaling of global environmental politics. *Annual Review of Environment and Resources*, 35, 255–282.

Arrow, K., Daily, G., Dasgupta, P., Levin, S., Mäler, K.-G., Starrett, D., *et al.* (2000). Managing ecosystem resources. *Environmental Science and Technology*, 34(8), 1401–1406.

Baumgartner, F. R., & Jones, B. D. (1991). Agenda dynamics and policy subsystems. *The Journal of Politics*, 53(4), 1044–1074.

Bergman, R. D. (2009). Life-cycle assessment of reused building materials: A proposed model for building professionals. *USDA Forest Service. Presented at* 2009 International Conference on Deconstruction, Building Materials Reuse, and Construction and Demolition Materials Recycling, Chicago. Available at http://www.bmra.org/S16Bergman.pdf.

Boughton, B., & Horvath, A. (2006). Environmental assessment of shredder residue management. *Resources, Conservation and Recycling*, 47(1), 1–25.

Bras-Klapwijk, R. M. (1998). Are life cycle assessments a threat to sound public policy making? *International Journal of Life Cycle Assessment*, 3(6), 333–342.

Braybrooke, D., & Lindblom, C. E. (1963). *A strategy of decision*. New York: The Free Press of Glencoe.

Briassoulis, H. (1989). Theoretical orientations in environmental planning: an inquiry into alternative approaches. *Environmental Management*, 13(4), 381–392.

Brulle, R. J. (2010). Politics and the environment. In K. T. Leicht and J. C. Jenkins (Eds.), *Handbook of Politics: State and Society in Global Perspective* (pp. 385–406). New York: Springer.

California Environmental Protection Agency. (2007). Life Cycle Assessment. *Department of Toxic Substance Control.* Retrieved April 13, 2010, at http://www.dtsc.ca.gov/TechnologyDevelopment/OPPTD_FLY_LCA.cfm.

Cohen, S. (2006). Understanding Environmental Policy. New York: Columbia University Press.

Cohen, S. (2011). The Transition from Environmental Politics to Sustainability Politics. *The Huffington Post,* January 31. http://www.huffingtonpost.com/steven-cohen/the-transition-from-envir_b_816198.html.

Cowell, S. J., Fairman, R., & Lofstedt, R. E. (2002). Use of risk assessment and life cycle assessment in decision making: a common policy research agenda. *Risk Analysis,* 22(5), 879–894.

Curran, M. A. (1997). Life-cycle based government policies: a survey. *International Journal of Life Cycle Assessment,* 2(1), 39–43.

Finnveden, G. (2000). On the limitations of life cycle assessment and environmental systems analysis tools in general. *International Journal of Life Cycle Assessment* 5(4), 229–238.

Finnveden, G., Hauschild, M. Z., Ekvall, T., Guine, J., Heijungs, R., Hellweg, S., *et al.* (2009). Recent developments in life cycle assessment. *Journal of Environmental Management,* 91(1), 1–21.

Gleick, P. H., *et al.* (2010). Climate change and the integrity of science. *Science,* 328(5979), 689–690.

Hammond, K. R., Mumpower, J., Dennis, R. L., Fitch, S., & Crumpacker, W. (1983). Fundamental obstacles to the use of scientific information in public policy making. *Technological Forecasting and Social Change,* 24(4), 287–297.

Harwood, J. (2011). An uphill battle on oil dependency. *New York Times, The Caucus* (Blog), June 12. http://thecaucus.blogs.nytimes.com/2011/06/12/one-alaskans-war-on-oil-dependency/?scp=8&sq=climate%20change%20nuclear&st=cse.

Holt, D., & Barkemeyer, R. (2010). Media coverage of sustainable development issues – attention cycles or punctuated equilibrium? *Sustainable Development,* 19. Doi: 10.1002/sd.460.

Hulse, C., & Herszenhorn, D. M. (2010). Democrats call off climate bill effort. *New York Times,* July 22. http://www.nytimes.com/2010/07/23/us/politics/23cong.html.

Jensen, A. A., Hoffman, L., Moller, B. T., Schmidt, A., Chriastensen, K., Elkington, J., *et al.* (1997). *Life cycle assessment: A guide to approaches, experiences and information sources* (Environment Issue Series No. 6). Denmark: European Environment Agency. Available at http://www.eea.europa.eu/publications/GH-07-97-595-EN-C.

Jones, C. O. (1974). Speculative augmentation in federal air pollution policy-making. *The Journal of Politics,* 36(2), 438–464.

Layzer, J. A. (2002). *The Environmental Case: Translating Values in to Policy.* Washington, DC: CQ Press.

Leinfellner, W. (ed.). (1990). *Uncertainty and Quality in Science for Policy.* Dordrecht: Kluwer Academic Publisher.

Leiserowitz, A., Maibach, E., Roser-Renouf, C., & Smith, N. (2011). Climate change in the American mind: Americans' global warming beliefs and attitudes in May 2011. *Yale University and George Mason University.* New Haven, CT: Yale Project on Climate Change Communication. http://environment.yale.edu/climate/files/ClimateBeliefsMay2011.pdf

Levin, K., Cashore, B., Bernstein, S., & Auld, G. (2007). Playing It Forward: Path Dependency, Progressive Incrementalism, and the "Super Wicked" Problem of Global Climate Change. Paper presented at International Studies Association Convention, Chicago, Il, February 28th – March 3.

Lindblom, C. E. (1979). Still muddling, not yet through. *Public Administration Review,* 39(6), 517–526.

Lowman, R. W. (1997). Life cycle assessment and public policy development for the automotive industry. Presented at *Total Life Cycle Conference and Exposition* 7–9 April 1997. Available at http://www.plastics-car.com/Recycling/Life-Cycle-Assessment-Public-Policy-Development/default.aspx.

Lyons, M. (1999). Political self-interest and U. S. environmental policy. *Natural Resources Journal,* 39(2), 271–294.

Nature [Editorial]. (2010). Climate of Fear. *Nature,* 464(7286), 141.

Norris, G. A. (2001). Integrating life cycle cost analysis and LCA. *International Journal of Life Cycle Assessment,* 6(2), 118–120.

Pizer, W. A., & Kopp, R. J. (2003). Summary and Analysis of McCain—Lieberman 'Climate Stewardship Act of 2003. S-139.' January 23. http://www.rff.org/News/Features/Documents/McCain-Lieberman.pdf (Retrieved January 25, 2011).

Rein, M., & Schön, D. (1993). Reframing policy discourse. In , F. Fischer and J. Forester (Eds.), *The Argumentative Turn in Policy Analysis and Planning* (pp. 145–166). Durham: Duke University Press.

Repetto, R. (2006). Introduction. In R. Repetto (Ed.), *Punctuated Equilibrium and the Dynamics of U. S. Environmental Policy* (pp. 1–23). New Haven: Yale University Press.

Revkin, A. C. (2011). Republicans get inconvenient replies at climate hearing. *New York Times, Dot Earth* (Blog), May 31. http://dotearth.blogs.

nytimes.com/2011/03/31/republicans-get-inconvenient-replies-at-climate-hearing/

Romm, J. (2009). Publicize or perish. *Physics World*, October. 22–23.

Ross, S., and Evans, D. (2002). Use of life cycle assessment in environmental management. *Environmental Management*, 29(1), 132–142.

Ruckelshaus, W. D. (1983). Science, risk, and public policy. *Science*, 221(4615), 1026–1028.

Schenck, R. (2009). The business case for life cycle assessment in US policy and legislation. *Institute for Environmental Research and Education*. Available at http://www.lcacenter.org/Data/Sites/1/SharedFiles/whitepapers/thebusinesscaseforlca.pdf Accessed on 8 June 2011.

Selin, H., & VanDeveer, S. D. (2011). Federalism, multilevel governance and climate change policy across the Atlantic. In P. F. Steinberg and S. D. VanDeveer (Eds.), *Comparative Environmental Politics*. Cambridge: MIT Press.

United Nations Framework Convention on Climate Change (UNFCCC) (2002). *A Guide to the Climate Change Convention and its Kyoto Protocol*. Bonn: UNFCCC Climate Change Secretariat.

U. S. Environmental Protection Agency (US EPA) (2000). Global Warming—Climate Change. Retrieved August 29, 2005, from http://yosemite.epa.gov/oar/globalwarming.nsf/content/climate.htm

U. S. Environmental Protection Agency (US EPA) (2000). *Framework for responsible environmental decision-making (FRED): Using life cycle assessment to evaluate preferability of products. National Risk Management Research Laboratory*. Washington, DC: Office of Research and Development.

U. S. Environmental Protection Agency (US EPA) (2001). *Life Cycle Engineering Guidelines*. Columbus, OH: Battle Columbus Laboratories.

U. S. Environmental Protection Agency (US EPA) (2006). *Life cycle assessment: Principles and practice. Prepared by* Scientific Applications International Corporation. *National Risk Management Research Laboratory*. Washington, DC: Office of Research and Development.

## Chapter 11

# Unexpected Appropriations of Technology and Life Cycle Analysis: Reframing Cradle-to-Grave Approaches

**Christopher Cummings, Jordan Frith, and David Berube**
*Department of Communication, Public Communication of Science and Technology (PCOST) Project, North Carolina State University*
dmberube@ncsu.edu

## Introduction

Technological adoption carries certain risks to human health and the environment. With the Industrial Revolution came new types of industrial injuries and disposal of potentially harmful waste. The growth of mass transportation in the early 19th century produced risks of disastrous accidents and toxic emissions at a previously unimaginable scale. In reaction to the undesired impacts of industry on human health and the environment a philosophy of industrial ecology emerged promoting foresight in technological advancement and adoption to maximize benefits and minimize risks to humanity and the environment. Explicitly, industrial ecology refers to the

*Emerging Technologies: Socio-Behavioral Life Cycle Approaches*
Edited by Nora Savage, Michael Gorman, and Anita Street

ISBN 978-981-4411-00-4 (Hardcover), 978-981-4411-01-1 (eBook)
www.panstanford.com

means and approaches of humanity to deliberately maintain a sustainable environment given ongoing technological advancement within ever-changing cultural and economic circumstances [1].

Industrial ecology is important yet difficult to research because the industry–environment interaction is complex. Each industrial action is linked to the actions of hundreds, if not thousands of other industrial agents, each with their own environmental impacts and consequences. At any given time an industrial agent may be developing, manufacturing, and disseminating their products to suppliers and consumers worldwide. Millions of individual products are shipped and sold to consumers, each with its own uses. Each product serves a primary purpose, however actual use may depart from the manufacturers' original intentions. Consumers may choose to use and dispose of products in drastically different ways than the manufacturers intended. Alternatively, they may choose not to purchase the products at all, meaning the products must be disposed by wholesalers. When considering product disposal, it is plausible for products to end up just about anywhere: deposited into landfills, littered along a roadside or within a waterway, or even incinerated [2].

## Industry–Environment Interactions

While the industry–environment interaction is most often mediated by individual consumer product use, there have been many examples of industry impacting the environment directly, and we have been shortsighted in considering the long-term consequences of such activity. The unpredictability of technological use has always been high. Our modern history is spotted with examples of industry-wide actions attempting to curb undesirable situations or fill desired needs, which in turn, create a myriad of unpredicted new problems. For instance, the rising threats of malaria and typhus in the 1940s and the growing concerns associated with vector control produced a concerted government policy. Dichloro-diphenyl-trichloroethane, more commonly known as DDT, was introduced for widespread use as an agricultural pesticide to kill mosquitoes around the world [3]. Unknown at the time, widespread use of DDT produced undesirable outcomes, including increased cancer rates and adverse effects on birds and mammals. These negative consequences eventually led to a ban on the use of DDT in the United States in 1972. A worldwide ban on its agricultural use was also initiated under the Stockholm Convention in 2001 [4].

Another example depicting the plague of detrimental shortsightedness in industrial impacts on the environment arose with the need for refrigerants. Refrigerators in the 1800s and early 1900s were reliant upon the use of several toxic gases including ammonia ($NH_3$), sulfur dioxide ($SO_2$), and methyl chloride ($CH_3Cl$). As the public demand increased for personal refrigeration units, there was a corresponding increase in fatal accidents and injuries due to inhalation associated with the leakage of the toxic gases or accidental incineration. An industrial opportunity arose for the manufacture and use of nontoxic and nonflammable refrigerants. In 1928, Thomas Midgley Jr. was credited with the discovery of a "miracle compound", a variety of chlorofluorocarbons (CFCs). The gases Midgley and his colleagues synthesized were later defined under the trademarked name "Freon" by DuPont [5]. Freon quickly replaced the widespread use of volatile gases for refrigeration worldwide. It wouldn't be until the late 1970s that concerns arose about the unintended effects of CFCs, primarily concerning stratospheric ozone depletion. Decades later the United Nations hosted multiple conventions and arranged multilateral environmental agreements including the Vienna Convention for the Protection of the Ozone Layer and the Montreal Protocol in attempts to respond to the new problems posed by industry's attempts to sate the needs of the bygone era. The potential threats of CFCs to the environment continue today [6].

In other instances, the impacts on human health and the environment are not caused by widespread industrial adoption of technologies but rather by consumers who may use products in unpredictable ways that jeopardize their health and surroundings. A recent and tragic example can be found in the Senegalese fishing town of Thiaroye Sur Mer, a suburb of Dakar. In Thiayore Sur Mer, the residents developed a process for extracting the lead from car batteries to fashion into weights for fishing lines to support the local fishing economy. To extract the lead, blacksmiths cracked open the car batteries and pulled out the lead. The widespread extraction and local disposal of lead made the soil in the town so poisoned with lead that animals and children began dying in significant numbers. The World Health Organization responded, and there are now plans to move entire neighborhoods from the town to hopefully curb the unforeseen associated dangers [7].

Examples like these call us to redirect environmental initiatives to better plan and facilitate the industry–environment interaction. The dominant approach to assessing potential industry harm to the environment may be the process of life cycle analysis (LCA). To maximize public and environmental health and safety LCA will need to find a way to include unanticipated consequences in its algorithm.

## Life Cycle Analysis

As a subset of industrial ecology, LCA attempts to understand potential environmental risks and has become a valuable tool in conceptualizing the industry–environment interaction. LCA has been formally conceptualized by two major organizations, the International Organization for Standards (ISO) and the Society of Environmental Toxicology and Chemistry (SETAC) [8,9]. The basic premise of LCA approaches is to assess the design, assemblage, dissemination, use, reuse, recycle of, and disposal of a material or technology—this has come to be known as a "cradle-to-grave" approach. The ISO defines LCA as: "A systematic set of procedures for compiling and examining the inputs and outputs of materials and energy and the associated environmental impacts directly attributable to the functioning of a product or service throughout its life cycle." The ISO also denotes a three-phase process for conducting LCA. First, an inventory of relevant system inputs and outputs must be compiled. Second, each input and output must be evaluated for potential environmental impact. Third, results of the first two phases must be interpreted and reported given the desired study objectives.

SETAC defines LCA as: "An objective process to evaluate the environmental burdens associated with a product, process or activity by identifying and quantifying energy and materials used and wastes released to the environment, to assess the impact of those energy and materials uses and releases on the environment and to evaluate and implement opportunities to affect environmental improvements. The assessment includes the entire life cycle of the product, process, or activity, encompassing extracting and processing raw materials;

manufacturing, transportation, and distribution; use/reuse/maintenance; recycling; and final disposal" [9]. Each definition highlights the scientific approaches of systematically defining parameters of study through operationalizing functional units of analysis and then assessing each unit individually. In this sense, LCA is an extension of traditional scientific methods that isolate units to be studied and then assess each unit rigorously in order to seek exploratory, and later, confirmatory data about technological impacts from cradle-to-grave.

The trouble with such attempted "objective" studies of technology is that technology is not an object that remains static and easily studied. Technologies are often defined by their materials, their form, and their components. This is a slippery slope that should be addressed. Technology must be defined by not only what it is and what it does, but also how it is used. A technology's instrumentality *in situ* often trumps its intended use. Consumers will use a technology as they see fit, regardless of the intended use of the technology. What lacks in the LCA approaches defined above is a social function that illustrates that technologies are constructions used in different ways by different people and to different ends. Because of the complexifying social factors involving technology use, the predictive capability of many industry–environment interactions is less than optimal. Efforts to quantify impacts using LCA are distorted by unpredictable uses of technology that render LCA as an overly idealistic and expensive tool that often provides underwhelming results.

It is the purpose of this chapter to challenge the underlying assumptions of LCA that do not account for the social construction of technology (SCOT) as a complexifying factor in cradle-to-grave assessment initiatives. In this chapter, we examine the use of silver nanoparticles in consumer products as a test case for the problems facing the future of LCA. We critique these approaches by drawing from both the consumer data on silver nanoparticles and the theory of SCOT that challenges the deterministic assumptions found in many LCA approaches. Silver nanoparticles are the most commonly used nanoparticles in consumer technology, and as they spread to more and more products, it becomes more difficult to perform robust LCAs on the potential risks of these particles. In the rest of this chapter, we briefly discuss the history of silver nanoparticles before moving on to our critique of current LCA approaches.

## Nano Silver

The toxicological signature and environmental fate of nano silver will not be debated below. For a comprehensive review of silver and nanotechnology, see Luoma's report [10]. While the scientific community does not downplay the potential utility of silver as an antibacterial, the full range of applications that have been advertised and/or marketed to the public in recent years has been extensive. In addition, how these products are used may affect issues of exposure and dosage that alter the hazardous potential to harm humans and the environment.

Silver ions and nanoparticles are used in brooms, food storage containers, drywall and paint, curtain coatings in hospitals, surfaces of all sorts such as surgical tools, wound treatments, and so forth. LG Electronics and Daewoo were selling nano silver product lines of refrigerators and vacuum cleaners. Sports companies such as Adidas, Polartec, Brooks Sports, and ARC Outdoors announced that they had intended and in some cases mounted silver particles as a disinfectant for clothing. Even a yoga company, Plank, claimed to sell silver particles as a disinfectant in their Cor soap. The existing research on silver nanoparticles focuses on very specific applications, mostly on the use of nano silver as a medical bactericide. Today, nano silver is being allegedly used in hair-straightening wands and curling irons, socks, towels, toothpaste, air purifiers, door locks, pet products, linen, tableware, water taps, baby blankets, etc. [11].

Silver is interesting for many reasons not the least of which is its various formulations at the nanoscale (a 9 nm nanoparticle has about 24,000 silver atoms in it): elemental (metallic), ionic (silver ion), nano silver (nanoparticulate silver), and colloidal [12]. Nowack, Krug, and Height in their comprehensive review of nano silver regulation revealed a clear example of how nomenclature and predicted use can complicate regulatory efforts. "Nano silver in the form of colloidal silver has been used for more than 100 years and has been registered as a biocidal material in the United States since 1954. Fifty-three percent of the U. S. Environmental Protection Agency (EPA)-registered biocidal silver products likely contain nano silver [13]. Most of these nano silver applications are silver-impregnated water filters, algaecides, and antimicrobial additives that do not claim to contain nanoparticles. Many human health standards for silver are based on an analysis of argyria occurrence (discoloration of the skin, a cosmetic condition) from the 1930s. The environmental standards on the other hand are based on ionic silver and may need to be re-evaluated based on recent findings that most silver in the environment, regardless of the original silver form, is present in the form of small clusters or nanoparticles" [14].

Silver's antibacterial property has been known for thousands of years and has been used for centuries in wound dressings. At the nanoscale, silver is appearing in an outrageously diverse set of applications if you are to believe all the reportage. Most of these products are making use of the antibacterial and antifuncidal properties of nano silver that are reportedly tested against more than 600 diseased organisms [15]. As nano silver moves into more and more consumer products, it may challenge LCA.

A classic example of this challenge is associated with the Samsung washing machine controversy in 2006. This story began with an article from Rhonda Rundle in the Wall Street Journal on June 6 [16]. Rundle claimed the EPA had performed an "investigation" of silver nanoparticles. According to the EPA's Office of Pesticides they had not and remained in the "early stages of developing our policies and approaches which are reflected in a draft white paper on nanotechnology." As such, the EPA's Office of Enforcement Compliance and Assistance either had not directed the Office of Pesticides on what steps EPA needed to take or they had and EPA was planning to make some announcement in the future.

The controversy over the Samsung washer seemed to have been precipitated by a request by NACWA (National Association of Clean Water Agencies). NACWA had this to say in the April 2006 issues of Clean Water News. "NACWA is closely tracking an emerging issue that will no doubt have implications for wastewater utilities in the future. The U.S. Environmental Protection Agency (EPA) has looked to nanotechnology for its potential to assist in environmental clean-up activities, but the unique properties that make nanomaterials effective at cleaning up pollution also make their behavior in the environment unpredictable" [17].

Silver maintains its powerful antimicrobial property at the nanoscale. NACWA learned of a new line of washing machines that relied on the disinfection properties of silver molecules and wrote to EPA's administrator, Stephen Johnson, to alert the agency to the potential impact these types of consumer products could have on water quality. Some strong studies indicate that silver may appear in waste streams and its effects on fresh water ecosystems are unpredictable [18–20].

The washer was purportedly designed to introduce silver ions and potentially particles into every load of wash. The impact of the silver discharges from these machines was apparently not a consideration when the machines was originally designed and approved for sale to consumers. Furthermore, whether these washing machines actually deliver antibacterial protection remained unclear [21]. Indeed, though Samsung stated that the washing machines involved nanotechnology, they subsequently removed that claim [22].

Nonetheless, NACWA expressed concern that nanotechnology may be one of those technologies that have outpaced our ability to evaluate potential long-term impacts. While many have looked to wastewater treatment plants to serve as the primary vehicle in removing these emerging contaminants from their liquid waste discharges, NACWA complained the roles source control and product substitution can play should these contaminants pose a threat to human health or the environment. TriTAC (the Technical Advisory Committee on State and Federal Regulatory Issues Affecting Publicly Owned Treatment Works represents three California Associations: League of California Cities, California Association of Sanitation Agencies, and the California Water Environment Association) made a similar claim [23].

The product line was called SilverCare and was advertised as nano in Australia and not so in the United States in 2005. Samsung's SilverCare Health System that included the washing machine of note refers to a suite of Samsung products including air-conditioning units and refrigerators [24]. Samsung's SilverCare washing machines were sold at Lowe's and Best Buy here and abroad. Now it seems to be called ActiveFresh.

The heart of the patented Silver Nano Health System technology of Samsung's washing machines was an electrolysis unit. Presumably, silver ions were extracted from a silver plate and delivered during the washing process into the washing water. According to Samsung, the silver nanoparticles were safe and eliminated as much as 99.9% of the harmful microorganisms, staph, and *Escherichia coli* [25].

Concerns expressed by the water-treatment claimants, centered about the claim that current treatments were able to remove only 50 to 90 percent of the silver. Friends of the Earth—Australia indicated worry about antibiotic resistance among harmful bacteria resulting from overexposure to silver in water [26]. Samsung claimed to have completed five toxicity studies on water fleas, fish, and earthworms and concluded that there was "minimal risk to the environment," but they did not share this research with the waste treatment companies, the press, or the public.

In addition, there was some concern the silver sludge may not be inactive. Sludge is commonly placed on farmland or in landfills according to Phil Bobel, a chairman from TriTAC. Samsung did claim the silver ions would quickly bind to organic matter and become inactive in waste. Benn and Westerhoff [27] warned that the amount of silver found in the sludge might limit its desirability as an agricultural fertilizer. While a secondary implication to the complaints, it remained an important concern to others especially when research indicates agricultural uptake of nano silver particles [28].

In 2006, the BUND (Friends of the Earth—Germany) began protesting Samsung's use of silver nanoparticles in their new line of washing machines. BUND argued that silver nanoparticles were not safe for consumers and that "animal experiments demonstrated unwanted effects of these minute particles" [29]. BUND encouraged Samsung to cease selling the washing machines, just as other consumer groups had successfully convinced Samsung to cease selling the product in Sweden in 2005 [30,31]. Such attention to the potential problems associated with the washing machines spurred a meeting between Samsung and BUND but did not bring about industry-wide change in product manufacture. The goal of BUND, to question the possible health risks associated with new technologies containing nanoparticles, is noble albeit their approach may seem more reactionary and combative than most would desire.

The problem addressed in the Samsung–BUND case is one that is common in environmental studies of silver nanoparticles: an emphasis on the particle rather than the technology in which it is present. BUND's argument was about the dangers of silver nanoparticles, but as these nanoparticles become ubiquitous in consumer products, it may not make sense to generalize about the risks of silver outside of the context within which consumers and the environment will come into contact with it. Instead, we need to recognize that silver nanoparticles may present different risks depending on the products and how consumers use the products. For example, the silver nanoparticles present in skin-care products or in the fabric of plush toys may not present the same human and environmental risks as those in water filtration systems.

Both NACWA and TriTAC called for something akin to LCA in their complaints to the EPA's Emerging Contaminants Workgroup. One of the responses recommended to the EPA by BNA's Phibbs-Rizzuto and others involved reclassifying the product as a co-pack. Originally, the machine was classified as a pesticide device and not a pesticide per se. However, the petitioners wanted it classified as a co-pack, meaning it would be both a pesticide and a pesticide device. That would mean the studies and tests required for marketing a new pesticide under FIFRA (Federal Insecticide Fungicide and Rodenticide Act) would need to be done. As a pesticide device there are some labeling, record keeping, and other requirements but much less onerous than as a pesticide [32,33].

Also of vital interest to the future employment of LCA methods considering nano silver may be the uncertainty surrounding its actual incorporation into consumer products. Nanoparticles of all types are themselves impossible for the average consumer to sense experientially through typical sensory processes like sight or touch. In addition, there is concern about the potential for fraudulent claims of nanotechnology use in consumer products. There also is concern about exaggerated claims of the sheer number of consumer products available for use today.

For example, the Project on Emerging Nanotechnologies (PEN) Consumer Product Inventory (CPI) has been the primary resource spurring overclaims by government, industry, non-government organizations, civic groups, media critics, and researchers about the number of nanoproducts on the market. The CPI has been touted as an excellent resource to track those consumer products that include nanoparticles and nanotechnology and are readily available for purchase. PCOST recently reviewed the CPI and tested the validity of the site through a content analysis of the "1000 + manufacturer-identified nanotechnology-based consumer products currently on the market" [34]. Our results were astonishing [35].

Findings from a content analysis and subsequent follow-up on actual product availability noted that only 54% ($n = 138$) of consumer products incorporating silver nanoparticles were actually available for consumer purchase and the majority of silver products (56%; $n = 145$) were only available in relatively few Asian countries including

China, Taiwan, Japan, Korea, and Singapore. We also noted from the study that most consumer products involving nanotechnologies were listed as business-to-business companies (39%; $n$ = 153) rather than as business-to-consumer companies (26%; $n$ = 103), and a substantial percentage of products were unlisted in their distribution channel (29%; $n$ = 113). From this study we concluded the "CPI is not wholly reliable, and does not have sufficient validity to justify its prominence as evidence for claims associated with the pervasiveness of nanotechnology on the U.S. and global markets. In addition, we caution researchers to approach the CPI with care and due consideration because using the CPI as a rhetorical flourish to amplify concerns about market intrusions seems unjustified" [11].

This study brings to the fore yet another complication for the future of LCA. Concerning nanoparticles like nano silver, there may be variance between what is purported by companies and what is actually being marketed to and purchased by consumers. Caution will need to be exercised to ensure that products claiming to incorporate nanoparticles actually do so, and that they are legitimate products that are readily available for consumption. Not being able to turn to groups like PEN for reliable information concerning product dissemination may hinder future LCA initiatives and make granular inventory assessment even more time consuming and expensive.

Next we turn to the SCOT theory of interpretive flexibility to show that even if we did have extensive and robust data on products containing nanoparticles, there is still no way to ensure that these products are used as intended. We then conclude this chapter by using our case study to make suggestions for the future of LCA.

## Social Construction of Technology

LCA has been the primary tool for conceptualizing the industry–environment interaction, but it is based on a fairly antiquated take on technology. LCA relies on a deterministic view of technology that places far too much control over the shaping of technology in the hands of the people who design and develop technologies. This "received view" of technology is present in many of the dominant technological determinism views that shaped the humanities and social sciences for much of the 20th century [35]. This conceptualization of technology promotes technologies as distinct from their social roles, functions, and interactions following product dissemination. Implicitly assumed within many scientific approaches to technologies in society, deterministic approaches view technology as something that comes from outside the social world to then affect the social world [36–39]. In other words, technologies are developed and released by manufacturers and then become autonomous in their effect on the social life of consumers. From this perspective, consumers have little control over what they do with a technology, how they use it, or what kinds of effects it has on society.

The central tenets of LCA share much in common with technological determinism. There is an implicit belief in cradle-to-grave analyses that technologies will be used for the purposes for which they are designed. In other words, they are released and then have their effect, and end-users have little control over how new technologies are shaped. The world of technological adoption is viewed as a closed system for which all variables can be accounted and controlled [40]. However, while deterministic views of technology dominated the social sciences and humanities for much of the twentieth century, they have been strongly criticized by theorists of technology. The attack on determinism began in earnest in the 1970s with the release of Raymond Williams' [36] *Television: Technology and Cultural Form*, and following Williams, a number of social constructivist approaches have risen refuting determinism, complicating the central tenets of cradle-to-grave LCA approaches that often missed these formative debates in the social sciences and humanities.

Emerging from science and technology studies, the most prominent attack on technological determinism may be the SCOT's theory [41–43]. SCOT, or technological constructivism, views technological adoption and use as a complicated matrix of interactions among designers, technologies, and consumers that often cannot be fully understood until after a technology has been introduced and used. To the proponents of SCOT, it is impossible to understand a technology without analyzing the social and cultural contexts in which it is embedded. One of the central tenets of many approaches to the SCOT is the concept of *interpretive flexibility*. Writing about interpretive flexibility, Pinch and Bijker [43] argue that technological artifacts are culturally constructed and interpreted; in other words, the interpretive flexibility of an object must be shown (p. 40). By this we mean not only that there is flexibility in how people think of or interpret artifacts but also that there is flexibility in how artifacts are designed. Interpretive flexibility allows SCOT researchers to examine both how cultural (and often strictly economic) values are built into the design of technologies and how cultural values shape how technologies are interpreted differently and used differently by social groups [44].

To get at a more concrete discussion of how end-users can often defy the intentions of designers, consequently complicating some of the core assumptions of cradle-to-grave LCA approaches, it helps to turn to Latour [45] and Akrich's [46] concept of *prescriptions*, which are "the behavior imposed back onto the human by non-human delegates [technologies]" [45] (p. 301). Technologies all have certain prescriptions built into their design that then influence how they are used. So, using Latour's famous example, an automatic door closer prescribes certain human behaviors (e.g., you cannot leave the door open, you must rush through the door). It is through the concept of prescription that Latour and Akrich develop the idea of machinic agency, and if they ended their exploration of technology with prescription, it would seem to be a useful boost to LCA proponents. After all, prescriptions encourage certain uses, meaning that it is possible to analyze these prescriptions and determine how a technology will be used. However, prescriptions are not fully deterministic; rather, prescriptions can only encourage behaviors. End users are always free, to some degree, to betray the prescriptions built into the technology. Returning briefly to the simple example of an automatic door closer, the user can always grab a rock and not allow the door to shut, defying the prescriptions built into the door.

A more detailed example of how this works is Akrich's examination of how technologies such as photoelectric lighting kits are often used very differently in Africa than designers expected. She acknowledges that designers do have a great deal of power over how a technology will be used, recognizing that some variables can be controlled in a cradle-to-grave analysis. However, despite the fact that "[d]esigners imagine and define actors with specific tastes, motives, prejudices {and} . . . *inscribe* this vision about the world and the users into the technical objects," end users often fail to gel with the designers' visions. They often betray the scripts built into technologies and "define quite different roles of their own" [46] (p. 208). So, for example, while designers built photoelectric lighting kits in a way to control when they could be used so they did not malfunction, these restrictions on behavior built into the scripts of the technology did not fit with the cultural values or the immediate needs of many of the African users. Consequently, they managed to bypass these restrictions, using the kits in ways not at all expected by the designers. This is a clear example of how difficult it is to predict how a technology will be used once end users begin rewriting the roles designers had attempted to define for them.

The unpredictability of end users has serious consequences for LCA future products containing silver nanoparticles. Even if all the problems discussed in the previous section are addressed and a robust and extensive dataset assembled, there is still no way to ensure that end users will play the roles assigned to them by designers. As silver nanoparticles become more and more ubiquitous in everyday products, this may become a serious problem. The difficulty in predicting how groups will adopt, use, and dispose of products ranging from washing machines to skin-care products is almost incomprehensible. At best, the analyst can make a guess as to how people will end up using these products. Best guesses will only get us so far, but they will continue to be all we have to rely upon until better assessment tools are established. The predictive quality of increased appropriations of silver nanotechnology products will diminish as the nanoparticles are incorporated into more and more products.

The consequences of interpretive flexibility can pose serious environmental risks. When end users betray the roles designed for them, they can endanger both themselves and others. Let's return to Senegal. The Senegalese example we mentioned earlier is a prime example of the consequences of users betraying their roles. To reiterate, in Thiaroye, a suburb of Dakar, the local people began extracting lead from car batteries to fashion homemade fishing equipment. The extraction process was dangerous and required blacksmiths to literally pull the lead from toxic, skin-burning acids that also leeched into soil. At least 18 children died from lead poisoning, and rates of miscarriage have escalated. Attempts to clean up the area have largely been unsuccessful [47].

The Thiaroye example demonstrates the interpretive flexibility of how serious the consequences can be from the use of fairly mundane technologies. For the Senegalese living in that town, the car battery served a secondary purpose as the primary source for fishing weights. The designers built prescriptions into car batteries that discouraged people from opening them, but that did not stop the people who saw them as a valuable source for lead. We are not criticizing the people of Thiaroye for defying the prescriptions; they likely had little idea of the consequences and were often faced with situations where these batteries were a viable source for fishing materials, and thus food and income. While this may be a somewhat extreme example of the flexibility of technology, it is nonetheless an important one. It serves as a tragic reminder that predicting the behaviors of end users is a task that involves so many variables that it is practically impossible. These examples do not suggest that LCA cannot be a valuable method of analysis, but they do show that we should be skeptical of any analysis that suggests that it can control for end user behavior as a variable.

Also, nano silver will be marked with unpredictable uses that discoverers and manufacturers may not intend. Intended product use may be neglected and consumers will trump attempts to corral technologic prescription, such as silver-impregnated clothing being repurposed for a toddler or cleaning sprays used to wash out kiddie bathing pools. It may not be long before nano silver may be used in ways and in places manufacturers and product designers actively discourage. For example, consumers often fail to read or dismiss warning labels addressing misuse.

Attempts to project the life cycle of nano silver will likely be shallow. Trying to assess where nano silver will be, whether it will remains nanosized or agglomerated into microunits, and how it will be taken up by organisms during and after anticipated and unanticipated consumer use are likely to remain unknown for many years, maybe forever. While some research is being completed to assess nano silver's environmental footprint, they remain unorganized and assume predictable uses. As such, there remain too few finite data inputs to assess its life cycle accurately.

While LCA approaches are complicated by social interactions mitigating the industry–environment relationship, we strongly urge its future use. We especially urge that future attention be paid to new social life cycle analysis (SLCA) methods that have been discussed in the last two years. SLCA attempts to incorporate some of the unpredictability into the analysis that we have discussed in this section. For example, SLCA acknowledges that with the production of technology, "social impacts are intrinsically related to the specific geographical and cultural context where they unfold" [48]. This approach attempts to include social-scientific and behavioral approaches in the ways it analyzes contexts of technological production and attempts to include culture as a variable. SLCA has gained recent attention in the LCA community, and we argue that more attention needs to be paid to these approaches.

However, SLCA may not go far enough. There is still a tendency in these approaches to focus on the site of production rather than the site of consumption. The quote above, for example, refers to the cultural context that affects production of technologies that result in analyses that mostly focus on regulation. What we are arguing for is distinct from such approaches; we argue that LCA must move in a direction where it is accepted as a valuable tool, but not relied upon as an objective measure that can capture all the complexity of technological adoption. Even quantitative approaches that gauge perceptions of certain technologies in certain cultures cannot predict how technologies will be used differently in different places [49], and how technologies being actually used will continue to be unpredictable.

While we argue that it will never be possible to control all the variables that affect technological adoption, LCA will nonetheless benefit from the incorporation of social-scientific and behavioral approaches that can examine sites of consumption and analyze how distinct groups use technology differently. These approaches will require fine-tuned, empirical approaches that avoid universal claims and instead rely on evidence drawn from specific sites. The goal in science (and much social science) is often to make generalizable, universal claims that can be applied to everyone and reproduced. For high-risk technological adoption, we argue that universal claims are not possible. There are too many variables involved in consumption to extrapolate research findings. However, just because unpredictability cannot be eliminated does not mean we should not try to control it as best as we can. Social-scientific approaches that examine specific sites of consumption will best be able to help us understand what people are doing with emerging technologies.

This chapter advocates caution about the unpredictable nature of technologic dissemination. Unlike uncertainty that is often quantified through statistical measures, unpredictability is the chaos that enters the otherwise knowable system. Uncertainty as a characteristic of study within LCA seeks the quantification of unknowns; it is the "known–unknown" quantity inherent in the industry–environment interaction. Some of the drawbacks of LCA approaches we argue are brought about by the unpredictable, the "unknown unknowns" that cannot be accounted in such foresight equations and algorithms. End users are one of those unknowns.

No approach can hope to predict all potential variables of influence. We are not criticizing LCA for its inability to do so; rather, we are criticizing the tendency to rely on LCA as if it is able to control for these forms of unpredictability. LCA must move away from deterministic approaches that negate the role of the end-user, and embrace unpredictable technologic use as an intervening factor in the complex industry–environment interaction.

## References


1. Duchini, F., & Hertwich, E. (2003). *Industrial ecology.* International Society for Ecological Economics. http://www.ecoeco.org/pdf/duchin.pdf. Accessed May, 25, 2011.
2. Graedel, T. E., & Allenby, B. R. (Eds.). (2003). *Industrial ecology*. New Jersey: Pearson.
3. World Health Organization. (1979). *Environmental health criteria 9: DDT and its derivatives.* www.inchem.org/documents/ehc/ehc/ehc009.htm. Accessed June 15, 2011.
4. Lear, L. (1997). *Rachel Carson: Witness for nature.* New York: Henry Hoyten.
5. DuPont (2011). *Freon®: 1930.* http://www2.dupont.com/Heritage/en_US/1930_dupont/1930_freon/1930_freon_indepth.html Accessed June, 15, 2011.
6. Environmental and Energy Study Institute (2009). *Chlorofluorocarbons: An overlooked climate threat.* www.eesi.org/100209_cfc . Accessed June 15, 2011.
7. World Health Organization (2008). *Senegal: Outbreak of lead intoxication in Thiaroye sur Mer.* http://www.who.int/environmental_health_emergencies/events/Senegal2008/en/index.html. Accessed June 15, 2011.
8. Environmental Management (2005). *ISO 14000;* CD-ROM format. Genève: International Organization for Standardization.
9. Baumann, H. (1996). 6th SETAC- Europe Meeting: LCA—Selected Papers. *International Journal of Life Cycle Assessment, 1*(3), 122.

10. Luoma, S. N. (2008). Silver nanotechnologies and the environment: Old problems or new challenges. Washington, DC: Project on Emerging Nanotechnologies. http://www.nanotechproject.org/process/assets/files/7036/nano_pen_15_final.pdf. Accessed June 29, 2011.

11. Berube, D. M., Searson, E. M., Morton, T. S., & Cummings, C. L. (2010). Project on Emerging Nanotechnologies—Consumer Product Inventory Evaluated. *Nanotechnology Law & Business, 7*(2), 152–163.

12. ICON (2008). Environmental Impacts of Nanosilver: An ICON Backgrounder. cohesion.rice.edu/centersandinst/icon/resources.cfm?doc_id=12722. Accessed December 20, 2012.

13. Fox News. (2007) Man claims skin treatment turned face permanent blue. Fox News.com. December 20.

14. Nowack, B., Krug, H. F., & Height, M. (2011). 120 Years of nanosilver history: Implications for policy makers. *Environmental Science & Technology, 45*, 1177–1183. Doi.org/10.1021/es103316q.

15. Powell, J. (1978). Our mightiest germ fighter. *Science Digest,* March, 57–60.

16. Rundle, R. L. (2006). This war against germs has a silver lining. Pittsburgh Post-Gazette. June 6. http://www.post-gazette.com/pg/06157/696127-114.stm. Accessed June 29, 2011.

17. Nanotechnology—Environmental boon or emerging pollutant? (2006). *Clean Water News*, April, 10–11.

18. Benn, T. M.. & Westerhoff, P. (2008). Nanoparticle silver released into water from commercially available sock fabrics. *Environmental Science & Technology, 42*(11), 4133–4139.

19. Blaser, S. A., Scheringer, M., *et al.* (2008). Estimation of cumulative aquatic exposure and risk due to silver: Contribution of nano-functionalized plastics and textiles. *Science of the Total Environment, 390*(2–3), 396–409.

20. Mueller, N. C., & Nowack, B. (2008). Exposure modeling of engineered nanoparticles in the environment. *Environmental Science & Technology, 42(*12), 4447–4453.

21. Samsung's Silver Nano washer ads reported exaggerated. (2005). *Appliance Magazine*, November 21. http://www.appliancemagazine.com/news.php?article=9434&zone=0&first=1. Accessed June 29, 2011.

22. EPA (2008). What you see is not always what it seems. *Pesticide Issues in the Works: Nanotechnology, the Science of Small* (July). http://

www.epa.gov/opp00001/about/intheworks/nanotechnology.htm. Accessed June 29, 2011.

23. Weir, C. (2006). Letter to Jim Jones, Director, Office of Pesticide Programs, US EPA. January 27.
24. Ubergizmo (2005). Samsung Silver Nano Health System. http://www.ubergizmo.com/2005/09/samsung-silver-nano-health-system/. Accessed December 20, 2012.
25. Uldrich, J. (2005). Nano is Samsung's silver living. Fool.com. December 23. http://www.fool.com/investing/high-growth/2005/12/23/nano-is-samsungs-silver-lining.aspx. Accessed June 29, 2011.
26. Senjen, R. (2007). Nanosilver—A threat to soil, water and human health? (March) http://nano.foe.org.au/sites/default/files/Nanosilver%20-%20a%20threat%20to%20soil,%20water%20and%20health%20March%202007.pdf. Accessed June 29, 2011.9
27. Benn, T. M., & Westerhoff, P. (2008). Nanoparticle silver released into water from commercially available sock fabrics. *Environmental Science & Technology, 42*(11), 4133–4139.
28. Harris, A. T., & Bali, R. (2007). On the formation and extent of uptake of silver nanoparticles by live plants. *Journal of Nanoparticle Research,10*(4), 691–695. Doi: 10.1007/s11051-007-9288-5.
29. Silver nano-technology in Samsung washing machines—statement on today's press release (2006). Brd-info.net. Original text: Samsung Electronics GmbH digital press briefcase: http://presseportal.de/story.htx?firmaid=14800 press briefcase via RSS: feed: /presseportal.de/rss/pm_14800.rss2.
30. Barwick, S. (2010). Colloidal silver banned in Europe, Part III: An exclusive interview with Anders Sultan, Sweden's largest colloidal silver manufacturer, Colloidal Silver Secrets. January 20. http://colloidalsilversecrets.blogspot.com/2010/01/colloidal-silver-banned-in-europe-part_20.html. Accessed June 29, 2011.
31. BUND warns against washing with nanoparticles. (2011). Eco Institut. http://www.eco-institut.de/e9579/e30975/e31799/news31866. Accessed June 29, 2011.
32. Bergeson & Campbell, P.C. (2006). EPA issues clarification regarding ion generating washing machines. November 29, http://www.lawbc.com/updates/112906-fifra.pdf. Accessed June 29, 2011.
33. EPA (2007). Pesticide registration; Clarification for ion generating equipment. September 21. http://www.epa.gov/fedrgstr/EPA-

PEST/2007/September/Day-21/p18591.htm. Accessed June 29, 2011.

34. Project on Emerging Nanotechnology, (2011). Consumer Products Inventory. http://www.nanotechproject.org/inventories. Accessed June 15, 2011.
35. Deloreto, J. (2010). Study: PEN Consumer Products Inventory Has "Substantive Deficiencies. " *NanoRegNews*. http://www.nanoregnews.com/article.php?id=415. Accessed June 15, 2011
36. Williams, R. (1975). *Television: Technology and Cultural Form*. New York: Schocken Books.
37. Veblen, T. (1963). The engineers and the price system. New York, NY: Harcourt.
38. Mcluhan, M. (1964). Understanding media: The extensions of man. New York, NY: McGraw-Hill.
39. Latour, B. (2005). Reassembling the social: An introduction to actor-network theory. Oxford, UK: Oxford University Press.
40. Slack, J. D., & Wise, J. M. (2005). Culture + technology : A primer. New York, NY: Peter Lang.
41. Feenberg, A. (2002). *Transforming technology: A critical theory revisited*. New York, NY: Oxford.
42. Bijker, W. E. (1992). The social construction of fluorescent lighting, or how an artifact was invented in its diffusion stage. In W. E. Bijker & J. Law (Eds.), *Shaping technology/building society: Studies in sociotechnical change* (pp. 75–104). Cambridge, MA: MIT Press.
43. Pinch, T. J., & Bijker, W. E. (1987). The social construction of facts and artifacts: Or how the sociology of science and the sociology of technology might benefit each other. In W. E. Bijker, T. P. Hughes, & Pinch T. J. (Eds.), *The social construction of technological systems: New directions in the sociology and history of technology* (pp. 17–50) Cambridge, MA: MIT Press.
44. Bowker, G. (1992). What's in a patent? In W. E. Bijker & J. Law (Eds.), *Shaping technology/building society: Studies in sociotechnical change* (pp. 53–75). Cambridge, MA: MIT Press.
45. Latour, B. (1988). Mixing humans and nonhumans together: The sociology of a door closer. *Social Problems, 35*, 298–310.
46. Akrich, M. (1992). The description of technical objects. In W. E. Bijker & J. Law (Eds.), *Shaping technology/building society: Studies in sociotechnical change*, (pp. 205–224). Cambridge, MA: MIT Press.

47. Lead for car batteries poisons an African town. (2009). *MSNBC.* http://www.msnbc.msn.com/id/28484477/ns/world_news-africa/t/lead-car-batteries-poisons-african-town/. Accessed June 15, 2011.

48. Zamagni, A., Amerighi, O., & Buttol, P. (2011). Strengths or bias in social LCA? *International Journal of Life Cycle Analysis.* Published Online. Doi: 10.1007/s11367-011-0309-3.

49. Assefa, G., & Frostell, B. (2007). Social sustainability and social acceptance in technology assessment: A case study of energy technologies. *Technology in Society, 29*, 63–78.

# Chapter 12

# Surface-Friendly Urban Mining

**Eric Lee Sander,**[a] **Charles William Edghill,**[a] **and Gary Roy Sander**[b]
[a] *Private Consultant*
[b] *Vehicle Reclamation Services of Maryland, LLC, 8019 Redstone Rd, Kingsville, MD 21087-1516, USA*
grsander.thorconsulting@att.net

## Introduction

The human population of the earth is projected to reach seven billion sometime this year (2011). The alarmists among us are predicting massive shortages of all sorts of natural resources as the demand for them grows.

Recognition of coming scarcities was one of the reasons many communities started recycling programs some years ago. Recyclables such as metal cans, bottles, and paper are collected one day a week and garbage another day. These programs weren't started only because of future scarcity but out of the realization that our landfills are reaching capacity in many areas. The availability of new places for landfills is difficult to find and, when found, is cost prohibitive. There is also revenue to be gained when selling the recycled material.

*Emerging Technologies: Socio-Behavioral Life Cycle Approaches*
Edited by Nora Savage, Michael Gorman, and Anita Street

ISBN 978-981-4411-00-4 (Hardcover), 978-981-4411-01-1 (eBook)
www.panstanford.com

We have been hearing about "peak oil" for a number of years but even the experts can't agree on when that will occur or if it has already happened. Even as working oil fields are depleting, there are known reserves that, for various reasons, are not allowed to be tapped. Oil from shale is another source that sits unusable at present.

A 1996 study by the U.S. Mineral Management Service of recoverable oil in Atlantic federal waters was estimated to be 7.2 billion barrels. The key word here is "estimated." It could be more or it could be less. Regardless, there are a lot of reserves that, if necessary, could be harvested. There is even more in Pacific waters.

That same study estimated there was 27 trillion cubic feet of natural gas available from these same Atlantic federal waters. Natural gas is also available from shale. It is presently being recovered in Pennsylvania from a 400-million-year-old rock formation called the Marcellus Shale. The U.S. Energy Information Administration estimates recoverable natural gas resources of 827 trillion cubic feet. The carbon footprint of natural gas is considerably lower than that of oil and coal. Some vehicles are already powered by it.

When thinking of metals such as steel, aluminum, and copper, realize that once mined and processed into finished products, the metal is still here even after the useful life of the product is over. Those three products are recycled and reprocessed many times after the initial mining. Think of it as "re-mining," or to use a term being more frequently used, urban mining.

As a species, humans have been very imaginative in creating things to meet our needs. Where we have been shortsighted is in designing those things so that they can be readily disassembled. Once a good/product has served out its useful life, it is destined for a junkyard or a landfill. Very little of the reusable material is removed and reused. There are over 4000 thousand airplanes that will never be flown again sitting in the desert in Arizona. Think of all the aluminum available there. If steel becomes in short supply there are 186,000 tons of Japanese shipping sunk in Truk lagoon. If we can raise the Russian nuclear submarine Kursk, which weighs 16,400 tons displacement from a depth of 360 feet, then salvaging those ships at Truk should be relatively easy if it should ever be desirable to do. How many tons of metals, plastics, and paper could be recovered from our landfills? This is radical thinking and probably will not be necessary for hundreds of years as there are more readily available commodities available through urban mining.

Another thing to keep in mind is that over 70% of the earth is covered by water. If equal amounts of the resources mentioned are there in the same quantities per square mile as we have found on land, depletion is delayed for many years. They may be more difficult to extract but humans have solved many problems in the past with far less knowledge and the technology we now possess.

We now have a proven, profitable example of urban mining in Europe. There are eight automobile disassembly plants in operation throughout the continent. A similar plant is planned to be built shortly in the United States. This country has an estimated 18 million end-of-life vehicles (ELVs) every year. Many wind up in open-air junkyards where they are pirated for parts. They are then sent to shredders where they are ripped into smaller pieces at a rate of 4 to 6 cars a minute. Magnets are then used to separate ferrous materials that can be reclaimed. What can't be salvaged then goes to a landfill where those unused materials add to water, soil, and sight pollution. All of these types of pollution are eliminated when the ELV is reprocessed as is presently being done in some facilities in Europe.

The reclamation and recycling of products/materials is not a recent phenomenon. For centuries, people reclaimed/recycled rags, paper metals, and other materials. Primarily due to low productivity in processing raw materials, the cost of making new products was too high to wantonly dispose of them as they neared the end of their useful lives.

Early in the industrial revolution, this changed. Unlike today, raw materials were readily available and easy to acquire at low cost from Mother Nature. Moreover, the productivity achieved through technological and process methodology changes caused the cost of processing the materials into new products to drop precipitously. Unfortunately, these productivity changes failed to be adopted by the reclamation/recycling industry. Therefore, when products reached the end of their useful lives their constituent raw materials were of so little value and the processing cost so high that they were disposed of unceremoniously in landfills, junkyards, and dumps. This view persisted for most of the last two centuries with the exception of the periods during the world wars and the great depression when the unavailability of virgin raw materials and economics made reclamation/recycling a necessity.

In 1970 the popularization of the environmental movement inaugurated a slowly increasing interest in reclamation and recycling.

Initially, in a desire to protect the environment and to protect us from being overwhelmed by our waste, many new laws and regulations were adopted that dictated the methods for the development of new sources of raw materials and the disposal of goods/products that have reached the end of their useful life. The net effect was a rise in the difficulty and cost of developing new resources and enhancing the relative economic attractiveness of reclamation and recycling.

Furthermore, today there is a growing Malthusian-style belief that the world is running out of even our most abundant natural resources. This belief is fueling a growing sentiment that reclamation and recycling of our waste is necessary to place our economy on a "sustainable" path.

Further in this chapter, the economic and social benefits of this type of operation will be discussed in detail. The depletion of resources on planet Earth may well happen at some future date but it will be further off than many now believe if urban mining becomes a common practice.

## The Life Cycle

Like people, the products we use in our modern economy have a life cycle. Beginning with conception/inception, each progresses through a period of growth/manufacture/construction, maturity/use/maintenance, death/end-of-economic-useful life followed by burial/deconstruction/demolition. Surface-Friendly Urban Mining™ involves both ends of the life cycle at the beginning, where raw materials are removed from the earth, and at the end, where the constituent materials are reclaimed from our waste to be recycled into new, useful products and goods.

Normally, virgin raw materials are not viewed as part of the urban-mining paradigm. However, they do set the standard by which the success of urban mining is measured. The cost and environmental impacts of acquiring and processing virgin raw materials establishes the price that can be achieved for the reclaimed and recycled counterpart. The bottom line in the success of urban mining is strictly defined by the economics and environmental advantages it provides over making goods and products from virgin raw materials.

Consequential life cycle analysis (CLCA) is a mathematical tool used to select the best strategy for the design, production, use, and disposal of a product that will maximize the benefits and minimize the costs to society. The strategy consists of the product design and the network of activities and associated processes that define the life cycle of the product from the extraction of raw materials from the earth used to make the product to materials processing, manufacture, distribution, use, repair and maintenance, reuse, and disposal or recycling.

Consequential life cycle analysis consists of a four step process:

(1) The first step is the definition of goals and scope. This is a statement of what is being studied and quantifies the service delivered by the product, the system boundaries, assumptions and limitations, allocation methods to partition environmental loads when several products or functions share the same process, and the impact categories chosen.

(2) The second step is an inventory analysis. This is the identification and quantification of the energy, resource use, and environmental releases associated with processes necessary to execute various activities associated with a particular design.

(3) The third step is the impact analysis. This is the technical qualitative and quantitative characterization of the impact of each of these activities on the environment, individuals, and society. Sometimes there may be uncertainty or disagreement on how to compute these impacts.

(4) The fourth step is improvement analysis. This is the evaluation of opportunities to vary the product design and network of activities and processes in the life cycle to maximize benefits and minimize costs not of just an activity but of the entire life cycle from-cradle-to-grave.

In terms of the life cycle analysis of lightweight vehicles, automotive disassembly is one opportunity to maximize benefits and minimize the economic, social, and environmental cost.

Improvement analysis implicitly assumes that it is possible to discern the best choice from a set of potential actions that might be taken. This can be difficult when some impacts are improved, but others of a different kind are worse. For example, the use of lightweight composite material may improve fuel efficiency of an automobile, but reduce the fraction of the vehicle that can be recycled.

This also can be difficult if different components of the community being served have different strategic objectives. Process and product developers are interested in CLCA as a way of incorporating environmental considerations into their design process. Consumers and consumer interest groups see CLCA as a way to better inform the customer of the relative environmental impacts of alternative products, hoping to bring to bear market pressures upon producers. Finally, regulators and policymakers see CLCA as a guide to environmental policy development and can provide a mechanism for enforcement of legislative objectives. It can be seen that these different components of our society can have very different and conflicting strategic objectives. No amount of analysis can directly resolve the conflict. As an alternative, analysis should be able to delineate the consequences of each alternative product design, process technology, or regulatory policy and provide a framework for negotiations. CLCA also brings up troubling issues of personal liberty when governments and self-appointed public interest groups use CLCA as a opportunity to intrude on the private affairs of private corporations in service of their own selfish agendas or even for purposes of extortion.

Traditional CLCA methodologies are concerned largely with the public health and environmental impact of the product life cycle. One premise of this chapter is that a more effective approach toward life cycle considerations for emerging ELV recycling technologies should include a wider range of perspectives and disciplines (e.g., economic and social). In particular, later on the chapter will examine environmental and other types of benefits and costs (e.g., economic and social) that should be evaluated in terms of recycling of automobiles and other products.

## Traditional Urban Mining

Traditional methods of urban mining are familiar to us all. The ubiquitous junkyards, landfills, dumps, and shredders permeate our communities. Even popular culture - TV, movies, songs, and art - have used their "image" to enhance the content of their art. It is the rare individual who doesn't have, somewhere, an image of no longer wanted goods stacked one upon another, slowly decaying as they wait to be eventually sent to the shredder where some portion - mostly the metals - can be eventually reclaimed and recycled.

Typically goods end up in the junkyard, landfill, or dump when, depending on the good/product in question, it has outlived its usefulness, requires repairs beyond its economic value, or is no longer fashionable. Initially, the goods are laid out or stacked up until such time as it becomes appropriate to deconstruct them. This may involve "cannibalizing" parts to use in extending the life of a similar good or shredding the good in order to reclaim the metals. Shredded pieces are then segregated by type of metal for sale to recyclers. The remainder of the good/product which consists, for the most part, of plastics, rubbers, glass, and dissimilar metals is difficult to separate. This shredder residue (SR) or "fluff," is usually relegated to a landfill. This SR can be a significant portion of the good/product.

As an example, in the traditional method for the urban mining of vehicles, processes focus almost totally on the recovery of metals. Comprising about 65% of the total weight of a vehicle, these metals - steel, aluminum, stainless steel, copper, brass, manganese, chrome, lead, and smaller amounts of numerous heavy metals - are eminently recyclable. Unfortunately, up to 20% of the metals are lost to the recycling process due to cross-contamination and the subsequent costs associated with sorting and separating the materials.

Shredded vehicles create approximately 500 pounds of non-metallic automotive shredder residue (ASR) per vehicle. This ASR is usually relegated to a landfill. The ASR accounts for approximately 35% of each vehicle and is expected to increase in the future as automobile manufacturers utilize greater percentages of lightweight materials. Eventually, about 5 billion pounds of "fluff" ends up in landfills every year, accounting for almost 4% of all landfill material.

This is becoming increasingly unacceptable to federal and state agencies, such as the local Departments of Solid Waste Management, and the EPA. These agencies are pushing hard to severely limit the practice of sending ASR to the landfill. But to accomplish this end necessitates a significant retooling of the reclamation/recycling process.

## The Assembly Line

On January 29, 1886, Karl Benz was awarded a patent for the automobile after a demonstration of the first viable horseless carriage. Over the next 25 plus years, many other famous inventors/

entrepreneurs followed suit. However, it wasn't until an ambitious and genius farm-boy turned machinist, turned entrepreneur from Michigan invented the assembly line in 1913, did the automobile progress beyond the status of a "rich man's toy."

Henry Ford's dissatisfaction with the quality and cost of the cars being produced, not only in his plants, and those of his competitors, drove him to search for a solution to the problem. Putting his genius to work, Ford rightly deduced that it wasn't the craftsmanship and drive of his workforce that was the problem; it was the basic nature of the manufacturing process.

Following the convention of the times, Ford initially utilized a methodology that has in recent years come to be called the batch and queue method. Basically it involves the fabrication of parts or the assembly of components in large lot sizes at separate and disparate work stations. Upon completion of the lot or batch they are transferred to a work-in-process inventory to await further assembly, ultimately, in this case into an automobile.

Beyond the scope of this discussion are the details of why the batch and queue manufacturing methodology is so unproductive. Suffice it to say, batch and queue introduces many challenges that defy good management. Quality, coordination, and communication issues are buried through high volumes of production and large inventories which in turn create their own special problems. Silos of responsibility emerge with each working toward their own ends rather than that of servicing the customer. Even with the application of sophisticated planning and control management techniques, the net effect is poor cash turnover, quality, communications, coordination, productivity, and deliveries.

Ford, in a stroke of sheer genius, revolutionized manufacturing through the creation of the assembly line, a concept currently known as "one-piece flow manufacturing." This methodology is based on the concept that the product - in Ford's case a car - moves on a conveying system from work station to work station with parts and assemblies being added at each until finally a completed product is achieved. In turn, each workstation is fed parts and assemblies by a line of workstations fabricating or assembling parts and assemblies - one part or assembly at a time.

Following this method, a pace or beat is established for each workstation to complete its work. The parts and assemblies must be, by necessity, "right" each and every time in order to not jeopardize the pace. In-process inventories become minimal at a level that insures the pace and balances the workload. Unnecessary processing steps, transportation, and motions are eliminated to enable and enhance the pace. As a result, productivity and quality skyrocketed to unprecedented levels. This enabled Ford to reduce the price of an automobile to a point where anyone with a good wage could afford to buy a Ford. Equally as important, he raised the wages of his workforce so that they too could enjoy the "delights of mother nature and freedom afforded by owning an automobile."

Since the introduction of the assembly line, other companies have adopted and refined Ford's concepts in building a plethora of different products. Most notably though is Toyota, which after the devastation of World War II needed to find a way to produce cars and small trucks at a low price and in small volumes. After studying Ford's methods for several months, Toyota's engineers, headed by Taiichi Ohno, developed what has come to be known as The Toyota Production System, or, in the United States as lean manufacturing. Today, companies all over the world are adopting the assembly line and lean techniques fathered by Ford and Ohno in order to provide high-quality and low-cost products and services to their customers.

## The Disassembly Line: Making Reclamation/Recycling Productive

Reclamation and recycling processes have not experienced the advances in productivity seen in the earlier stages of the life cycle, particularly the manufacturing stage. Reclamation of materials from our waste streams has always been viewed as a dirty business unsuited for, and unglamorous to, process geniuses such as Ford. Hence the reclamation industry has wallowed in an unproductive swamp. Fortunately, this is in the process of changing.

In the late 1980s, two Dutch gentlemen, Mr. Frans van den Mosselaar and Leo Th. M. Reuser, one a long-time reclamation entrepreneur and the other a banker, came together with the dream of creating a new reclamation process. A disruptive process that, by its very nature, enables the effective and productive reclamation and recycling of all of a good's/product's constituent materials.

Many attempts have been made by the major automobile producers such as Ford, General Motors, Chrysler, BMW, Mercedes

Benz, and Volvo to change the urban-mining paradigm. Individually and together they have spent tens of millions of dollars building large complicated and highly automated/computerized disassembly lines. Unfortunately, they failed miserably due to their line's inability to handle more than a limited number of makes and models of vehicles (normally just one make and model). Frans and Leo on the other hand saw that the solution necessitated small distributed facilities utilizing a semi-automated disassembly line that placed an emphasis on the flexibility of the workforce to adapt to virtually any vehicle. Since 1991 Frans and Leo, operating as Autorecycling B.V., have dismantled hundreds of thousands of vehicles in their plant in Dongen, The Netherlands. None of their competitors have achieved this goal and none are still in business. So what makes them different and thus successful?

The disassembly line as designed by Frans and Leo is an adaptation of Ford's assembly line - just running in reverse. Consisting of 15 workstations, as adapted for use in the United States, the vehicle - car or light truck - proceeds through a series of activities in which various parts are removed until at the end all that remains is an empty steel hulk that is then baled into a 24″ × 24″× 30″ steel "log."

The first four workstations involve the preparation of the vehicle for disassembly. During this stage the vehicle identification number is checked to ensure it was not stolen, the "owner" is paid, and the vehicle is "greened." All fluids, fuel tank, engine computer, battery, mercury switches, and air bags are responsibly removed to avoid environmental and safety issues. The title is removed from registry with the Motor Vehicle Administration to ensure the vehicle will never return to the road, and the vehicle is evaluated to determine how to best dismantle it (Fig. 12.1).

The remaining 11 workstations involve the active dismantling process. In order to enter the disassembly process, the tires and wheels are removed and the vehicle is placed onto the conveying line. Primary disassembly involves the removal of the doors, glass, door coverings, and seals. Secondary disassembly furthers the process by stripping away the bumpers, rear/headlight assemblies, trunk lids, composite fenders, seats, Boston board, speakers, high-up brake light, interior computer, and seat coverings. At the third workstation, the interior is removed including the carpet, headliner, side panels, console, seat belts, visors, radios/speakers/DVD players, mirrors, GPS, interior lights, window cranks, and dashboard.

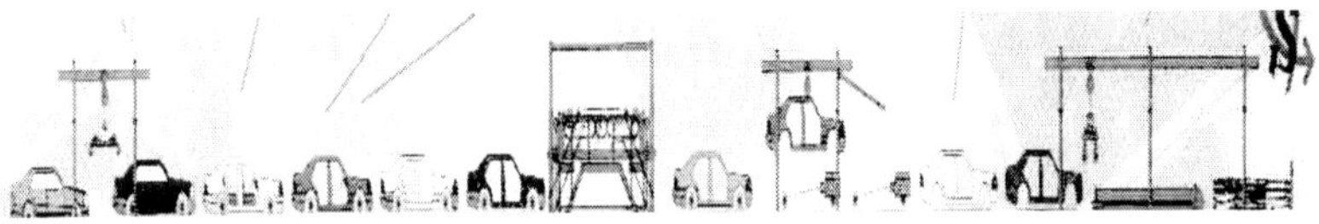

| WC# | Description |
|---|---|
| −4 | Vehicle Receiving |
| −3 | Greening |
| −2 | De-Titling |
| −1 | Triage |
| 0 | Line Entry |
| 1 | Primary Disassembly |
| 2 | Secondary Disassembly |
| 3 | Interior Disassembly |
| 4 | Car Inverter |
| 5 | Drive Train Disassembly & Hoist |
| 6 | Electrical Disassembly |
| 7 | Final Harvesting |
| 8 | Hulk Evaluation |
| 9 | Pre-Prep |
| 10 | Compaction |

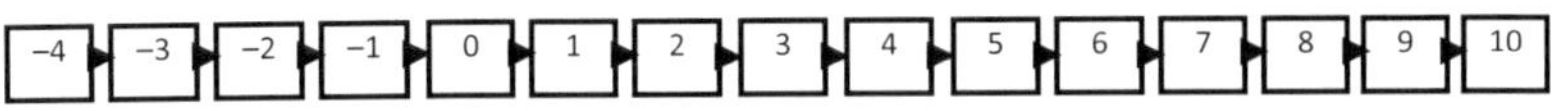


**Figure 12.1** Car recycling systems B.V. disassembly line workstation definition.

The fourth workstation differs from the others in that the vehicle is upturned 180 degrees by a hydraulically-driven inverter. This is done to provide comfortable and safe access to the underside of the vehicle. The disassembly specialist is then able to remove the exhaust pipes, muffler, catalytic converter, rear shocks, and brake lines before cutting the front shock towers and cross-members that support/secure the engine and finally up righting the vehicle. In the fifth workstation the drive train and body are separated along with the radiators, suspension, steering, and servo-assist units. Then the body is lifted and moved into the sixth workstation where the electrical harnesses, motors, horns, and heater/air conditioner unit(s) are removed. Final harvesting of the vehicle takes place in the seventh workstation for items either missed in the earlier workstations due to the time constraints of the beat or as a result of being unique to the vehicle. The eighth and ninth workstations involve preparation of the hulk for baling in the tenth workstation as dictated by the size of the vehicle or as by customer specifications.

Using this carefully orchestrated and performed methodology, 99.5% of the vehicle can be reclaimed with near zero cross-contamination. This level of performance provides multiple benefits to the operators of the disassembly lines, customers, and society as a whole.

Having an approximately 50% greater yield per car over the traditional junkyard shredder, in addition to a 40,000 vehicles per annum production capacity advantage, provides the operators the ability to generate much greater income. First of all, operators will have high volumes of items such as reclaimed plastics, glass, and rubber that were previously sent to landfills, a great attraction to offer customers and a benefit for which they will be willing to pay a premium. Second, premium prices will be generated as a result of the operator's ability to provide high volumes of quality materials to customers who then will develop new products that would not be viable if the volume of quality materials wasn't available. Third, low contamination/high quality products reduce the number of steps necessary to process and, therefore, reduce the cost of recycling the materials, an advantage that will entice customers to pay more. Fourth, by reducing processing steps, the amount of energy used and the subsequent reduction in the amount of carbon dioxide released into the environment are attractive benefits to a customer and society. Fifth, significantly lower levels of waste produced by the disassembly line means less has to be sent and buried in landfills, thereby, reducing disassembly operation's disposal fees.

## Automotive Urban Mining/Disassembly in America

An automotive disassembly plant will soon be built near Havre de Grace in Harford County, Maryland, by Vehicle Reclamation Services of Maryland, LCC (VRS MD) (Fig. 12.2). This urban mining facility, using just-in-time logistics, is located inside an ordinary looking building and does not look like a junkyard.

**Figure 12.2** The automotive disassembly plant.

This facility will be the first, in the United States, of a series of state-of-the-art indoor, environmentally benign, semi-automated reclamation operations (Fig. 12.3). The facilities will be based upon a proven Dutch-manufactured disassembly system, capable of processing thousands of end-of-life automobiles and small trucks per annum and producing huge concentrations of reclaimed materials and parts. This facility will have the disassembly capacity of 167 vehicles per day, or 40,000 vehicles per year, 20,000 for each of the two disassembly lines. There are eight facilities that already exist in Europe.

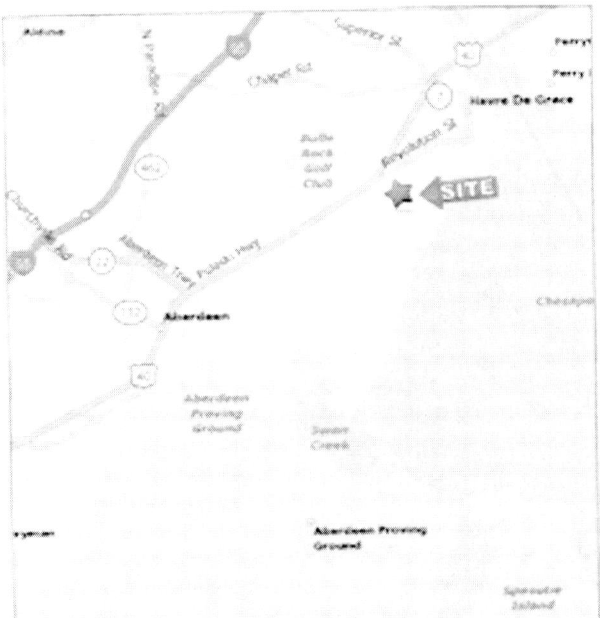


**Figure 12.3** Location of the automotive disassembly plant.

Upon arrival, ELVs are immediately placed on the disassembly line. When disassembled, the resulting parts, automotive products, and materials are immediately shipped off to waiting customers providing a one-stop source of feedstock for manufacturers, materials and parts remanufacturers, and reclaimed parts retailers with pre-existing orders - everything except the honk. Sales for the most part are executed under prenegotiated term (1 year) and long-term (3 to 5 years) purchase agreements. The recovery of more of the valuable materials makes the recovery of the less valuable materials that would otherwise be uneconomic to recover more worthwhile.

Annually, in the United States alone, about 18 million automobiles are available for reclamation, most of which pass through traditional, open-air shredder/junkyards that have significant environmental and regulatory problems. There is also a significant reduction in land use requirements formerly necessary for landfill disposition. VRS MD will be certified "environmentally responsible" by the ARA (Automotive Recyclers Association) and will recover enough fuel from the recycled cars to generate their own electricity to operate the entire facility.

Due to the wide variety of automobile makes, models, and customized design changes by the owners, it takes a semi-automated system that uses a highly-skilled, well-paid, local labor to imaginatively operate, maintain, and even redesign facility equipment and processes to handle changing conditions. It is just not practical to ship old automobiles long distances to obtain cheap labor - which soon becomes expensive labor once trained - especially with increasing fuel and transportation costs in the future. Also, it is uneconomic to expose skilled and expensive labor to dangerous processes that cause frequent injuries or contact with toxic substances. Considerations such as this have been historically of great importance for improving worker remuneration and working conditions.

Parts and material recovery makes an ELV an asset that can be sold to a disassembly facility, especially if the owner recovers a deposit paid at the time of purchase. It is far better to recycle here than to abandon the car or send it off to a foreign country where after a short life it is abandoned or poorly reclaimed and recycled using methodologies dangerous to workers. In addition, the last thing we want to do is export our pollution.

Missed by many is the understanding that prices and profits are reflective of the value the customer places on the seller's products. For customers and society as a whole, the benefits they derive from new and less expensive/high quality products, accompanied by a more sustainable, cleaner environment, will allow the urban miner to enjoy high profits which will provide, at mature operational capacity, a combined eight-figure tax revenue for federal, state, and local governments from this facility alone for years to come. The plan is to build over 200 of these operations in the United States.

From where do the automobiles come? Companies with automobile disposition requirements are finding fewer solutions to their needs. Auto insurance companies, fleet-leasing companies, rental car companies, major auto manufacturers, municipalities, and others need efficient, environmentally sound, legal, and high volume disposition facilities. For example, GMC produces approximately 20,000 "lemon" cars a year and is without a satisfactory disposition program for these vehicles. In addition, future regulations may require automobile manufactures to provide the means for end-of-life, cradle-to-grave programs for the vehicles they manufacture. In the future, cars may actually be designed to be recycled. Manufactures will want to maintain the materials stock in the economy to suppress materials costs. Also, insurance companies sell their older low-valued collision vehicles at auction but the returns (less than 8% of Actual Cash Value) have been falling dramatically in recent years due to liability concerns about the final disposition of the vehicles. Insurance companies need to defray the cost of claims and be assured that the vehicles are permanently off the road.

Volume buyers of reclaimed original equipment manufacturer (EOM) automotive materials and parts fall into several categories: R&D companies, materials manufacturers, parts remanufacturers, export parts buyers, export automobile rebuilders, and reclaimed parts retailers. This is just a partial representation of the markets. The market demand for reclaimed automotive materials and parts is worldwide, exceeds supply, and is constantly growing.

With support from strategic partners, additional operations can and will be put in place throughout the United States. VRS MD conservatively estimates, supported by studies from Ernst & Young and General Motors, 100 plants operating around the country would generate more than $4.5 billion in revenue. Furthermore, for every five plants, ancillary businesses will be established involving automotive core recovery and distribution, glass processing, aluminum processing, wholesaling of automotive parts, processing/wholesaling of engine coolant, bumper assembles, air-bag assembles, and brain-box assembles. In addition alternative fuel products will be produced through pyrolysis processing of tires, rubber and plastics.

## What Are the Materials that Are Reclaimed?

In citing the opportunities available to earn profits, and provide benefits to customers and society as a whole, it must be emphasized again that these benefits accrue from large volumes of quality product. So how much can be expected to be harvested from a single vehicle, disassembly plant and from all the anticipated plants?

Tables 12.1 and 12.2 show the expected recovery by generic type of product, part, or material.

**Table 12.1** Vehicle parts and material

| Reclaimed item or package* | Weighted average units per vehicle† | Explanation |
|---|---|---|
| Ferrous package (R.M.) | 2000 lbs | Average vehicle yield 2000 lbs |
| Non-ferrous package (R.M.) | 40 lbs | Copper/alum./alloy/lead, Average vehicle yield 40 lbs |
| Foam (R.M.) | 20 lbs | From seats, Average vehicle yield 20 lbs |
| Agglomerated plastics/rubber (R.M.) | 25 lbs | Lenses, trim, gaskets, bumpers, etc. Average vehicle yield 25 lbs |
| Oil / Hydraulic fluids (R.M.) | 1.25 gallons | From engines, transmissions, brakes, etc. Average vehicle yield 1.25 gallons |
| Engine coolant (R.M.) | 1 gallon | From engine and radiator average vehicle yield 1 gallon |
| CFC refrigerant(R.M.) | 1 lb | From air-conditioning system average vehicle yield 1 lbs |
| Tires (R.C.P.) | 3 tires | 3 tires average yield per vehicle |

| Reclaimed item or package* | Weighted average units per vehicle† | Explanation |
|---|---|---|
| Spare tire assembly (R.C.P.) | 1 | 1 tire assembly average yield per vehicle |
| Wheels ( steel ) (R.C.P.) | 2 wheels | 2 steel wheels average yield per vehicle |
| Wheels ( alloy ) (R.C.P.) | 1 | 1 alloy wheels average yield per vehicle |
| Hubcap (R.C.P.) | 3 | 2 hub caps average yield per vehicle |
| Gas tank (R.C.P.) | 1 | Steel or plastic0.75 gas tanks average yield per vehicle |
| Radiator (R.C.P.) | 1 | Copper or aluminum 1 radiator average yield per vehicle |
| Catalytic converter (R.M.) | 1 | Rare metals and case 1 Catalytic Converter yield per vehicle |
| Battery (R.C.P.) | 1 | 1 battery yield per vehicle |
| Air bag (R.C.P.) | 2 | 2 air bags average yield per vehicle |
| Eng.-tans assembly(R.C.P.) | 1 assembly | 1 engine average yield per vehicle 0.75 for remanufactured |
| Rack and pinions (R.C.P.) | 1 assembly | 1 assembly average yield per vehicle 0.75 for remanufactured |
| Core package (R.C.P.) | 1 core package | Starters, alternators, brake shoes, rotors, etc. (core package) |
| Body parts (R.C.P.) | 1 body part package | Doors, hoods, trunk lids, bumpers, taillights. (body parts package) |
| Glass (R.M.) | 1 | Wind shield and doors |
| Electronic systems (R.C.P.) | 1 unit | Radios, speakers, GPS, DVD players |

*These weighted numbers represent an average recovery spread over many processed vehicles.

†The Reclaimed Materials (R.M.) and Reclaimed Component Parts (R.C.P.) are either logged, baled, bundled, barreled, boxed, caged, containerized, or palletized and sold by the ton, pound, hundred-weight, hundred-count, cage, or container load.

*Note:* The fuel (gas/diesel) reclaimed from each vehicle is approximately 5 to 8 gallons and will be used to generate electricity for the facility. This will accrue to $7/vehicle energy cost avoidance for the operation.

**Table 12.2** Potential for reclaiming materials

| Product | Per vehicle yield | Annual per plant yield | Annual 400 plant yield |
|---|---|---|---|
| Steel | 1 ton | 40,000 tons | 16,000,00 tons |
| Aluminum and copper alloys | 40 lbs–0.02 tons | 1,600,000 lbs – 800 tons | 640,000,000 lbs – 320,000 tons |
| Plastics/rubber | 25 lbs | 1,000,000 lbs – 500 tons | 400,000,000 lbs – 200,000 tons |
| Oil equivalence of plastics rubber @4.25 lbs/gal | 5.88 gals | 235,200 gals – 5,600 bbs | 47,058,820 gals – 2,240,000 bbs |
| Engine blocks | 0.95 | 38,000 | 15,200,000 |
| Transmissions | 0.95 | 38,000 | 15,200,000 |
| Gasoline | 8 gals | 320,000 gals – 7620 bbs | 128,000,000 gals – 3,047,620 bbs |
| Oil | 1.25 gals | 50,000 gals – 1,190 bbs | 10,000,000 gals – 238,095 bbs |
| Antifreeze | 1 gal | 40,000 gals – 952 bbs | 16,000,000 gals – 380,800 bbs |
| Freon | 1 lb | 40,000 lbs | 16,000,000 lbs |
| Windshield washer fluid | 0.5 gals | 20,000 gals – 476 bbs | 4,000,000 gals – 190,476 bbs |

## Urban Mining: Beyond the Automobile

Automobiles are not the only candidates for the application of the disassembly process. Any hard good/product that man has produced is available and capable of being reduced to its basic components. Everything that consists of parts and made of different materials from white goods (household appliances) to electronics, lawn-care machines, recreation/exercise machines/equipment, houses/buildings, furniture, airplanes, and ships/boats can be dismantled using a disassembly line.

Just as with the automobile, the disassembly line can provide high volumes of and significantly improved yields in reclaimed materials when compared with the traditional method. But, there is another very important reason for adopting the disassembly methodology. For some items such as electronics and batteries that are constructed in part of highly toxic compounds/elements, the traditional methods expose workers and the environment to harmful if not deadly fumes and particulates. Presently, this problem is managed by shipping the potentially dangerous products to third world and other countries that have little concern for their people and environment. In addition to this total disregard for the wellbeing of these people, the cost of transportation of goods/products to the point of reclamation and eventually to the customer that will recycle the reclaimed materials generally offsets the benefit of this effort.

Urban mining can greatly reduce the need for toxic and unsavory mining operations driven by the demand for cell phone. For example, the cell phone or smart phone contains materials like copper, gold, lead, nickel, antimony, zinc, beryllium, tantalum, mercury, arsenic, and coltan, among others. The plastic shell contains material crude oil and natural gas. Coltan is a vital element in the production of cell phone–level capacitors, and the majority of it is mined in the Democratic Republic of Congo with deplorable working conditions and it is claimed that local gorillas are hunted to provide food for the workers. The United Nations estimate that 60% of the world's population owns a cell phone. Some people are on their second, third, or fourth cell phone. Only about 10% to 15% of cell phones are recycled in the United States. In many countries recycling is virtually unknown.

## Design for Reclamation/Recycling

Beginning in the 1980s, much effort was placed on the design of goods/products to enable and enhance more effective reclamation/recycling. This culminated in legislative guidelines in both Europe and North America. Much of this focus has been on the ubiquitous automobile, though other products - particularly electronics - have come under great scrutiny as they have become such a large portion of modern life.

In the United States the big three big automobile companies, under the auspices of the U.S. Council for Automotive Research (USCAR), Detroit, spend a lot of time on the recyclability of vehicles. USCAR's Vehicle Recycling Partnership along with the Recycling Materials

Center, a program of the Center for Environmental Policy, Economics and Science (CEPES) and in conjunction with dismantlers, have been working together to find the best ways to enhance the recyclability of automobiles. In Europe the Consortium for Automotive Recycling (CARE), consisting of 10 automobile companies [BMW (GB) Ltd, Fiat Auto SpA, Ford/Jaguar, Mercedes Benz (U.K.) Ltd, Renault (U.K.) Ltd, VAG (U.K.) Ltd, Vauxhall Volvo Cars (U.K.), PSA (Peugeot /Citroen)], is a collaborative project of vehicle manufacturers and vehicle dismantlers aimed at reducing the amount of a vehicle's weight going to landfill.

The issues these groups have established as the greatest challenges facing the group and hindering progress toward a more recyclable automobile are fasteners and materials.

Fasteners hold the automobile together. Whether they are nuts, bolts, clips, connectors, lugs, snaps, zippers, Velcro, adhesive, or any of a multitude of different devices, all need to be strong and in sufficient amount/numbers to assure a safe, rattle-free, and reliable vehicle. First and foremost, the design of the vehicle needs to provide the customer the most value for his money - this means that any compromise that has to be made has to fall in favor of providing strength and security.

Automobile manufacturers face this issue both as a part of the construction/assembly of the vehicle and, in reverse, the disassembly/recycling effort. By in large, the number/type of fasteners, dictates the amount of time and labor that is spent in assembling of an automobile. This, of course, affects their cost and profitability. Many innovative fasteners have been developed over the last 125 years to solve this labor/cost issue with remarkable results – cars are better, more reliable, of higher quality, and safer today than ever before. As a result, the connector's strength results in connections that are less likely to fail when struck by a shredder's hammer. In cases where similar materials are involved, this is of little concern. Unfortunately, if the still connected materials are dissimilar the expense of isolating and separating the pieces skyrocket the labor expense and/or add to the amount of ASR generated by the traditional shredder process. As more dissimilar materials are introduced, more ASR will be generated. Over 20 years have been dedicated to this effort and many innovations have been developed with only marginal success. Maybe changing the type and number is not the solution – changing the disassembly process is.

Since the oil embargo in 1973, the automotive industry has been under great pressure from governments to enhance fuel mileage and reduce polluting exhaust emissions. Automobile engineers have accomplished much to address these issues. By the introduction of electronic engine controls, engineers have doubled the thermodynamic efficiency of internal combustion engines while simultaneously reducing "harmful" pollution to levels unbelievable in the 1970s. Furthermore, automotive design has evolved from being as aerodynamic as the side of a barn to a level that has reached the limits dictated by both safety and passenger comfort. To make the vehicle lighter, new designs are much smaller than their 1970s counterpart and significantly lighter as the result of the use of "new" materials.

Steel, a marvelous material because of its inherent strength, durability, malleability and cost, has long been the foremost component of the automobile. Unfortunately, it is heavy. Therefore, to make vehicles lighter, the engineers have introduced other materials where some of steel's attributes are unnecessary and use of the proper, lighter substitute is appropriate. The dash, interior panels, fluid reservoirs, hubcaps, and even fenders are now of plastic. Likewise, aluminum has been introduced to fenders, suspension parts, engines, transmission cases, and numerous other applications that employ properly selected alloys and designs. Other materials such as magnesium and stainless steel are potential contenders for providing alternatives to steel; however, their costs, without a significant benefit in strength or manufacturability (stainless steel, that weighs more than carbon steel, does provide corrosion resistance at high purchase and workability costs), hinder their effectiveness.

Why does the use of more alternative materials impact recyclability?

- More dissimilar materials necessitate more connections and more ASR.
- Many of the alternative materials – thermoplastics, aluminum, copper – cannot be easily reprocessed for their original use.
- Some alternative materials contain harmful elements (lead, mercury, cadmium, and hexavalent chromium) that need to be handled carefully when recycling. This is particularly true as a result of the addition of multiple electronic devises in our newer vehicles.

Designing for recycling is difficult. It involves many compromises that can impact the appeal, longevity, integrity, safety, operating cost, fuel efficiency, emissions, and cost of the good/product. If the goal of recycling is to extend the lifetime in which scarce materials are available, the solution appears to be improvement of the disassembly process and extension of the useful life of the good/product.

## Importance of Recycling to Human Civilization

We would like to end this chapter by presenting the results of a first-order analysis of the long-run importance of recycling through urban mining to human civilization. The 21st century will become the first century of the age of scarcity. The world's rapidly growing industrial complex has a voracious appetite for raw materials, but raw materials are becoming increasingly more expensive and difficult to obtain from conventional mines. These mines that are often located in far away, unfriendly, or unstable countries with poor human rights, and environmental protection records require more and more energy and increased carbon footprint to process poorer ores and produce severe toxic chemical pollution. Constraints in the near future on the availability of fossil fuels to power mining operations and processing, land-use restrictions, and environmental laws may limit access to virgin sources. One method to mitigate this problem is urban mining, a highly efficient method of recycling to obtain prime-quality, uncontaminated, used materials from secure and well-regulated domestic sources, thereby, delaying the exhaustion of virgin raw materials. Urban mining is an important complement to renewable energy for establishing a more sustainable society.

A sustainable society is one that meets the needs of the present without compromising the ability of future generations to meet their own needs. To become materially and energy sustainable, the economy's throughputs would have to meet Herman Daly's conditions:

- Its rates of use of renewable resources do not exceed the rates of regeneration.
- Its rates of use of nonrenewable resources do not exceed the rates at which sustainable renewable substitutes are developed.

- Its rates of pollution emission do not exceed the assimilative capacity of the environment.

The second condition implies that we should delay the exhaustion of nonrenewable resources for as long as possible. The best way to do this is through recycling.

Virtually all metals can be recycled into high-quality new metal. The process varies for different metals but generally produces metals of equivalent quality. Worldwide, over 400 million metric tons of metal is recycled each year. There are also other environmental benefits, for example, using recycled steel to make new steel enables the following reductions compared with processing virgin materials:

- 86% in air pollution
- 40% in water use
- 76% in water pollution

Metal recycling is a major worldwide industry. Manufacturing of metals continues to be one of the largest United Kingdom manufacturing sectors, employing more people, and contributing more value to the U.K. economy, than motor and aerospace combined. The U.K. is one of the five largest metal-scrap exporting countries in the world. Metal recycling is a £5 billion U.K. industry, processing ferrous and nonferrous metal scrap into vital secondary raw material for the smelting of new metals. The industry employs over 8000 people and makes a net contribution to the U.K. balance of trade. Thus recycled metals have significant economic value and so scrap metal is rarely discarded or sent to landfills (Table 12.3).

Recycling of materials obtained from urban mining requires considerably less energy than processing raw ore. Carbon dioxide and toxic emissions are similarly reduced.

**Table 12.3** Energy savings from recycling

| Recycling scorecard | New metals made using recycled metals | Energy savings |
|---|---|---|
| Aluminum | 39% | 95% |
| Copper | 32% | 86% |
| Lead | 74% | 60% |
| Steel | 42% | 62–74% |
| Zinc | 20% | 60% |

*Source*: Reproduced with permission from British Metal Recycling Association (BMRA)

A recently submitted (January 2011) European Commission staff working document on the prevention and recycling of waste indicates that end-of-life vehicles (ELVs) can be a significant source of recycled materials if existing legislation is fully implemented compared with construction and demolition waste (C&D) and packaging and electrical and electronic equipment waste (WEEE) (Table 12.4).

**Table 12.4** ELVs can be a significant source of recycled materials

| Material (MT) | C&D | ELV | Packaging | WEEE | Total |
|---|---|---|---|---|---|
| Metal | 15,684,560 | 4,294,855 | 3,626,182 | 1,571,160 | 25,176,757 |
| Ferrous metal | 15,684,560 | 3,842,765 | 2,822,122 | 1,428,327 | 23,777,774 |
| Nonferrous | – | 3,452,090 | 804,059 | 142,833 | 1,398,982 |
| Aluminum | – | – | 633,057 | – | 633,057 |
| Glass | – | 169,534 | 11,941,037 | – | 12,110,571 |
| Aggregates | 329,375,760 | – | – | – | 329,375,760 |
| Plastic | 7,842,280 | 565,113 | 5,367,396 | 714,164 | 14,488,952 |
| Paper | – | – | 30,548,745 | – | 30,548,745 |
| Wood | 15,684,560 | – | 4,924,692 | – | 20,609,252 |
| Total | 368,587,160 | 5,029,501 | 56,408,052 | 2,285,323 | 432,310,037 |

*Source*: http://ec.europa.eu/environment/natres/pdf/Resource_Efficiency_Final.pdf (copyright European Commission, 2011)

Another study analyzes recycling rates, recycling contents, and share of old scrap in the total scrap flow for 60 metals at global level (Table 12.5). It shows that end-of-life recycling rates are still globally low due to the relative abundance of primary material and due to the absence of performing collection and processing of old metals. Recycling rates are above 50% only for 18 metals of 60 and the share of old scrap in the total flow is above 50% only for 13 metals. Metal recycling could be improved in the future and contribute to meet a significant part of future demand. To that end, collection and processing of old metals should be improved as well as the design of the products to increase the "recyclability" of some metals. Notice in the table below that the global rate of recycling of lithium is less than 1%.

not look like a junkyard. This facility will be the first in the United States of a series of state-of-the-art indoor, environmentally benign, semi-automated reclamation operations, capable of processing thousands of end-of-life automobiles and small trucks per annum, and producing huge concentrations of reclaimed materials and parts. This facility will have the disassembly capacity of 167 vehicles per day, or 40,000 vehicles per year for each of two disassembly lines. A combined eight-figure tax revenue for federal, state, and local governments can come from this facility alone for years to come. VRS MD conservatively estimates that 100 plants operating around the country would generate more than $4.5 billion in revenue. The plan is to build over 200 of these operations in the United States alone.

Parts and material recovery makes an ELV an asset that can be sold to a disassembly facility, especially if the owner recovers a deposit paid at the time of purchase. Auto insurance companies, fleet-leasing companies, rental car companies, major auto manufacturers, municipalities, and others need efficient, environmentally sound, legal and high volume disposition facilities. Volume buyers of reclaimed EOM automotive materials and parts fall into several categories: R&D companies, materials manufacturers, parts remanufacturers, export parts buyers, export automobile rebuilders, and reclaimed parts retailers.

Due to the wide variety of automobile makes, models, and customized design changes by the owners, it takes a semi-automated system that uses a highly skilled, well-paid, local labor to imaginatively operate, maintain, and even redesign facility equipment and processes to handle changing conditions. It is just not practical to ship old automobiles long distances to obtain cheap labor—which soon becomes expensive labor once trained—especially with increasing fuel and transportation costs in the future.

Annually, in the United States alone, about 16 to 20 million automobiles are available for reclamation, most of which pass through traditional, open-air shredder/junkyards that are unsightly and have significant environmental and regulatory problems. Automobile shredder residue is usually relegated to a landfill. The ASR accounts for approximately 35% of each vehicle, and is expected to increase in the future as automobile manufacturers utilize greater percentages of lightweight materials. With automotive disassembly, there is a significant reduction of ASR and landfill requirements. The process also will recover enough fuel from the recycled cars to generate their own electricity to operate the entire facility. The recovery of more of the valuable materials makes the recovery of less valuable materials economically viable.

It is far better to recycle here than to abandon the car or send it off to a foreign country where after a short life it is abandoned or poorly reclaimed and recycled using methodologies dangerous to workers. In addition, the last thing we want to do is export our pollution.

Automobiles are not the only candidates for the application of the disassembly process. Everything that consists of parts and made of different materials from white goods (household appliances) to electronics, lawn-care machines, recreation/exercise machines/equipment, houses/buildings, furniture, airplanes, and ships/boats can be dismantled using a disassembly line.

To obtain a successful transition to a more sustainable society, we should delay the exhaustion of nonrenewable resources for as long as possible. The best way to do this is through recycling. Automotive disassembly can produce a significant fraction of the recycling for society and maximize recycling efficiency for ELVs. Calculations show that a good recycling program can significantly increase supply, reduce price, delay exhaustion of the total stock of nonrenewable materials hundreds of years, and provide more time for transition to the use of renewable sources. The sooner we begin a serious recycling program, the larger the available stock of nonrenewable resources in the future. Because of inadequate recycling efficiency in the past, a significant fraction of our initial resource of many materials is already lost.

There is an urgent need for action because the 21st century will become the first century of the age of scarcity when the world's rapidly growing industrial complex has a voracious appetite for raw materials, but raw materials are becoming increasing more expensive and difficult to obtain from conventional mines often located in far-away, unfriendly, unstable, and unsavory countries and requiring more and more energy and increased carbon footprint to process poorer ores and producing severe toxic chemical pollution. Constraints in the near future on the availability of fossil fuels to power mining operations and processing, land-use restrictions, and environmental laws may also limit access to virgin sources

# Dedication

This chapter is dedicated to Bill Hyman, "the Junkyard Dog," who brought the automotive disassembly process to the United States and also to Roy Sander who had an eye for a business opportunity and liked to talk to strangers—even one with a pig tail, tie-died shirt, wearing sandals and jeans and introduced him to us.

# Coda

Intensive exploration and technological improvements may increase the amount of raw materials we can extract from the ground, but they will only delay and not prevent eventual exhaustion of these virgin resources. As we proceed through the age of scarcity, urban mining will become the dominant and then the only paradigm for obtaining supplies of minerals and other raw, non-renewable resources that form the basis of our industrial civilization. Automotive disassembly and, more generally, product disassembly will be an important methodology for obtaining high-quality, non-contaminated raw materials. Because recycling efficiency can never reach one hundred percent, all nonrenewable resources will eventually disappear—dispersed through the environment in such low density to make them uneconomic to recover—an entropy death. We must then either tap non-terrestrial sources to delay this evil day or do without.

## References

Aluminum Industry Statistics web site www.aluminum.org.

American Metals Market web site www.amm.com.

Automotive Recycling Association web site www.a-r-a.org.

Batties, Stephanie J., & Adler, Robert K. Production, energy, and carbon emissions: A data profile of the iron and steel industry, www.eia.doe.gov.

British Metal Recycling Association (BMRA) web site: http://www.recyclemetals.org/

Car Design On line, www.cardesignonline.com/production/end-of-life-recycling.php

Car Recycling Systems B.V. website www.crs-europe.com

Control and reduction of environmentally hazardous substances. Available at http://www.globaldenso.com/en/csr/report/2011/documents/e16.pdf

Development of high speed multispectral imaging for sorting automotive plastic, National Recovery Technologies, Inc. 566 Mainstream Drive, Suite 300, Nashville, TN 37228-1223, 615-734-6400

End-of-Life Vehicle Solutions website elvsolutions.org

Environmental guidelines for aluminum manufacturing, *Aluminum Manufacturing Magazine,* 317–322.

European Commission (2011). Commission staff working document accompanying the communication from the Commission to the European Parliament, the Council, the European Economic and Social Committee and the Committee of the Regions on the thematic strategy on the prevention and recycling of waste, SEC(2011) 70 final, Brussels, le 19.1.2011 http://eur-lex.europa.eu/LexUriServ/LexUriServ.do?uri=SEC:2011:0070:FIN:EN:PDF

Forbes Magazine article about William W. Hyman entitled Junkyard Dog (Copy can be provided on request)

Graedel, T. E, Allwood, J., Birat, J.–P., Reck, B. K., Sibley, S. F., Sonnerman, G. Buchert, M., Hagelüken, C., *Recycling Rates of Metals: A Status Report,* second report of the Global Metal Flows Working Group of the International Panel on Sustainable Resource Management of UNEP (2011). Available at http://www.unep.org/resourcepanel/Portals/24102/PDFs/Metals_Recycling_Rates_110412-1.pdf

Gross, Daniel (2007). The Tao of junk, *Slate Magazine,* September 8. Available at http://www.slate.com/articles/business/moneybox/2007/09/the_tao_of_junk.html

Institute for Local Self Reliance website www.ilsr.org

Institute of Scrap Recycling Industries, Inc., www.isri.org/misi5-prod/isri/-government-relations/design-for-recycling.aspx?hkey=53669ac2-2e2b-4a7c-9865-if4bb6oe9d50

Lean Enterprise Institute website www.lean.org

Meadows D., Randers J., & Meadows, D. (2004). *Limits to Growth, The 30 Year Update* (p. 338). Vermont: Chelsea Green Publishing Company.

Plastics Industry Association SPI website www.plasticsindustry.org

Product stewardship opportunities within the automotive industry, prepared for the Minnesota Office of Environmental Assistance under a grant from U.S. EPA & Region 5, prepared by Five Winds International, www.fivewinds.com

RecycleinMe website www.recycleinme.com

Recyclenet Spot Market Pricing for Recycled Materials website www.scrapindex.com

Recycling of metals, University of Cambridge Web site: http://www.doitpoms.ac.uk/tlplib/recycling-metals/index.php

Reduce, reuse and recycle, *BMW Magazine* (Copy provided on request)

Roper, L. David, *Depletion Theory*, Department of Physics, Virginia Polytechnic Institute and State University, Blacksburg, Virginia 24060-0435, roperld@vt.edu, 1976. http://www.roperld.com/science/minerals/DepletTh.pdf

Roper, L. David, Lithium depletion including recycling, http://www.roperld.com/science/minerals/lithium.htm

Steel Development Company website www.steeldevelopment.com

Steel scrap recycling drops as reserves decline (actually about how there is a shortage in steel scrap), *Metal Producing and Processing,* August, 2007.

Transportation Technology R&D Center, Argonne and industry to tackle end-of-life vehicle recycling, www.transportation.anl.gov/features/recycling_crada.html

U.S. Geological Survey (2009). Mineral commodity summaries 2009, p. 195. http://minerals.usgs.gov/minerals/pubs/mcs/2009/mcs2009.pdf.

U.S. Geological Survey (2011). Mineral commodity summaries 2011, p. 98. http://minerals.usgs.gov/minerals/pubs/mcs/2011/mcs2011.pdf.

Wilt, C., & Kincaid, L. (1997). There auto be a law: End-of-life vehicle recycling polices in 21 countries, *Resource Recycling* (March issue).

# Conclusions and Suggestions for Future Work

In the introduction, we sketched a vision of a life cycle analysis (LCA) that would incorporate the traditional methodology but also encompass the way in which technologies are embedded in social contexts and with the transformations created by emerging technologies. The two vignettes illustrate social and behavioral impacts of novel technologies that would not be covered by traditional LCA.

Traditional LCA works mainly on what Allenby (Chapter 3) calls Level I, where the focus is on the materials that go into a product, the energy used in manufacturing, and the waste that result—including their toxic and environmental risks. At Level 2, the local system in which the technology is deployed by human beings should be considered. Level 3 adds the impact on the overall global system. Ramachandran (Chapter 5) further explains the limitations of traditional LCAs for nanomaterials and technologies, illustrating an important aspect of Allenby's Level 3—the difficulty involved in anticipating the potential global transformations of emerging technological systems.

Traditional LCA also involves assessing whether or not a compound presents toxicity, and if so to what extent. Toxicological testing and high-throughput screening are relatively easy to perform. The problem lies not with the performance of these tests, not with the resulting information and data provided, but rather with the ecosystem of that data. Understanding the toxicity of a compound is only one facet of the challenge. The question which we need answer to is what is the compound encountered at the point of exposure. Clearly it is not the pure, simple compound. This conundrum is not evinced anywhere as clearly and distinctly as it is when encountering nanomaterials whose alteration of shape, size, functionalization, *etc.* results in changes to resulting data obtained from toxicity tests (see Ramachandran, Chapter 5).

As populations increase and our ecosystem becomes more burdened, polluted, and strained, societies tend to focus on the "contaminant of the day" or the current "politically correct" term. This action in itself would not be so dangerous and self-destructive were it not for the fact that activities are undertaken simultaneously, which are either patently empty and void of any true impact or are even more deleterious than the one for which a remedy is sought. The development of proactive environmental approaches along with anticipating and preventing adverse ecological impacts is needed. There is little question about the need, the debate lies in the path to achieve the end goal.

One path is to perform expensive, time-consuming toxicity tests on single compounds and then develop models to predict toxicity and impacts upon cells, tissues, organs, and organisms. This strategy, however, with the possible exception of improvements in model development is neither new nor the best tool for understanding ecological impacts. With the occurrence of multiple compounds reacting with each other and other substances (synergistic effects) in the environment, the transformation of compounds as materials are employed (intended and non-intended usages), recycling, and disposal, along with the various bio-geo-chemical reactions taking place in the ecosystem, render this approach simplistic at best, misleading at worst.

Another approach is to use LCA to track and mark the environmental impacts through the analysis of a material or product throughout its life cycle stages. While more comprehensive in terms of moving beyond single compounds, this approach also falls a bit short. There is little or no consideration of social science — behavior, cultural norms, societal attitudes. In addition, it is not clear how to extrapolate this data such that it is applicable for complex mixtures (true ecological exposures) or toward the development of anticipatory approaches.

The authors in this book have proffered ways of filling in these missing pieces and moving toward more holistic ecological protection. These ideas are innovative, yet not impossible; they are striking, yet not outlandish; they are functional, yet not foolproof.

The authors have, herein, initiated a dialogue with, and the editors hereby invite you to read, analyze, consider, and contribute to this dialogue. Among the alternatives is the option of forming another organization and forging ahead with meetings, workshops, and symposia. Another option is to continue business as usual and plod along behind the toxicologists assessing the toxicity of materials, seek remedial solutions, and stand watch as cancer rates continue to increase, as more species become extinct, and as the environment deteriorates.

The continued reliance on "numbers" or notified data obtained from toxicity tests, because of the ease and familiarity with numerical data, lulls us into a false sense of security. Armed with numbers, we seek to protect the ecosystem and the reality is this approach is failing. Embracing and incorporating social sciences and qualitative data can provide a better approach. Not a cleaner one, not a simpler one, but one with improved results. The addition of economic factors into environmental considerations was also difficult. Yet its incorporation produced a more holistic approach. Indeed, as the divide between the very rich and the very poor increases, as the middle class continues to shrink, perhaps whether or not a product is economically beneficial for an industry or society should not override the social and cultural demands. The reliance on toxicity and economic data to the exclusion or reduction of other important parameters has not proven successful. Cancers and other environmentally related public health threats are increasing, novel technologies heralded to improve the quality of life often do not reach those who are truly in need of them most, and assaults on the environment continue. It is time for a different tactic, for a more holistic approach.

Several of the chapters describe extensions to LCA that incorporate social, economic, and political factors. Ramachandran emphasizes the importance of considering the cognitive heuristics used by the expert. Social life cycle analysis (SLCA), mentioned by Cummings *et al.* (Chapter 11), emphasizes the way in which technologies are social constructs. Additionally, consequential life cycle analysis (CLCA), as applied by Sanders *et al.* (Chapter 12), considers the environmental and economic benefits of urban mining for civilization in the light of dwindling resources. Cohen (Chapter 10) emphasizes the importance of incorporating the political and economic contexts into LCA-type analyses. Marchant (Chapter 4) adds the importance of the regulatory context

One method for integrating the social and behavioral contexts in LCAs is to add stakeholders to the process of developing, analyzing, and implementing LCAs, in effect bringing proxies for the social directly into the LCA. Ellwood *et al.* (Chapter 6) recommend including entrepreneurs, scientists, engineers, and social scientists into the LCA process. Pauwels (Chapter 8) gives an example of a trading zone that conduct an LCA across disciplinary and epistemic boundaries. Both Ellwood and Pauwels recognize that an extended LCA involves consideration of ethics and the values of different stakeholders.

What is needed is a comprehensive capability that uses LCA as a model but extends beyond current versions ways that allow us to adaptively manage emerging socio-technical systems like Allenby's 'Five Horsemen' (nano, bio, info, cogno, robo). We propose a life cycle scenario analysis (LCSA). The interactions among the 'Five Horsemen' and multiple social systems are unpredictable but not entirely beyond the anticipation of a core group that could include the sorts of expertise in this volume, plus the kind of public engagement expertise that exists in the Centers for Nanotechnology in Society.

The goal of the core LCSA group would be to come up with several scenarios for how these socio-technical systems might develop, including which ones were more or less desirable from the standpoint of different stakeholders. Each scenario would have at least very general qualitative and quantitative metrics attached to it. How do we know if we are heading into one of the scenarios — or if the emerging systems are going in a direction not anticipated at all?

Within the larger LCSA, LCAs could be designed to consider parts of these emerging systems (e.g., the invention, production, and retirement of a technology of a particular nano-biotechnology designed to remove arterial blockages (see Chapter 2). If such a technology were not feasible or had high risks, then it would not have to be considered in the higher-level scenarios involving emerging technologies and human health.

The first step would be to pilot an LCSA on the 'Five Horsemen', documenting the process by which the group worked as well as all of their ideas and drafts leading up to the resultant scenarios. The pilot could begin with just a core group to see if they could develop scenarios, including planning what other stakeholders should be consulted, and any other LCAs, including perhaps SLCAs and

CLCAs, that should be done in coordination with the LCS. One end result would be scoping out a budget for a serious and deep LCSA that would include a long-term adaptive management strategy for altering the scenarios as new information emerged and changing the composition of the core group over time. Eventually, ongoing LCSAs might span generations.

# Index